W9-BQS-824

ACTIVITY SOLDER
Tip Tinner
THERMALTRONICS
www.thermaltronics.com
ZIPPY
COMPACT
1000
amazon
RECTORSEAL
EP-200 Epoxy Putty
Sets Hard As Steel in 15 - 20 Minutes
WORKER
WORKER
N-STRIKE
ELITE
MODULUS
ACCESS
STR
TURNIGY
COMPACT CHARGER
2S / 3S
Direct AC in
Cells Equalizer
1S
2S
3S
NERF
NERF
6
FAST SET
Clear, medium
bodied solvent
cement

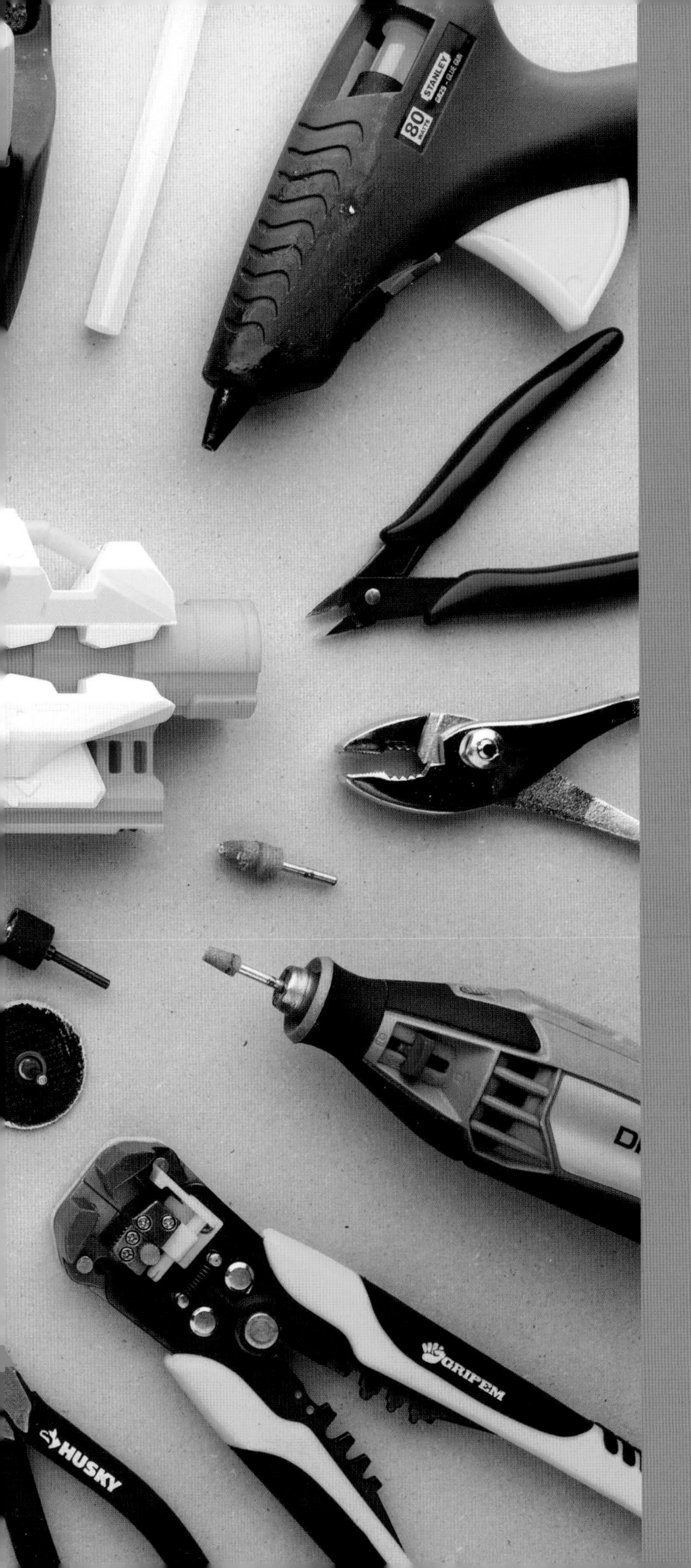

THE

NERF BLASTER MODIFICATION GUIDE

THE UNOFFICIAL HANDBOOK FOR MAKING YOUR FOAM ARSENAL EVEN MORE AWESOME

LUKE GOODMAN

Contents

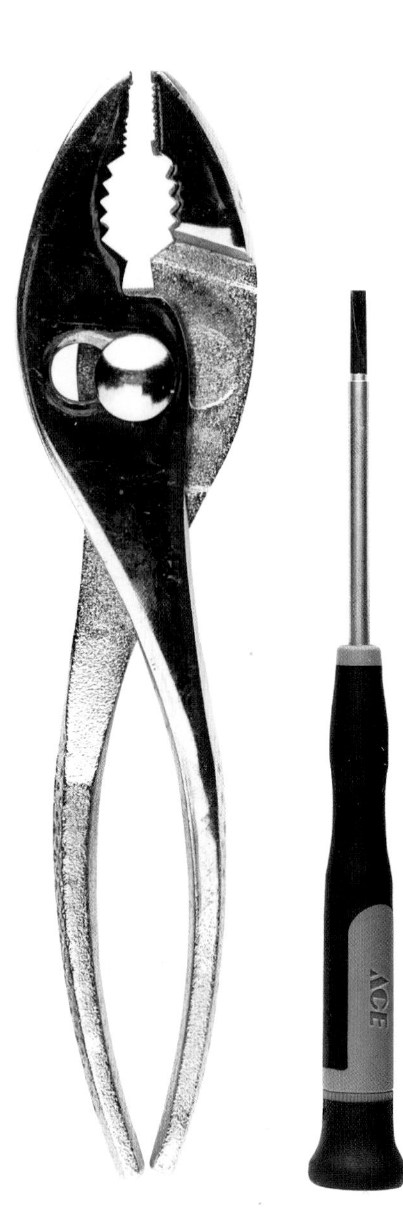

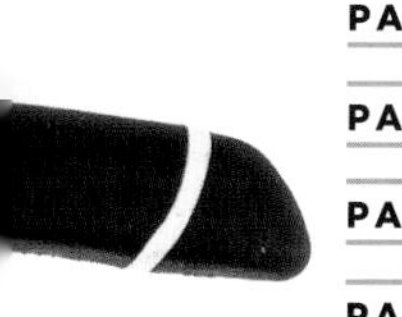

STRYFE
NERF

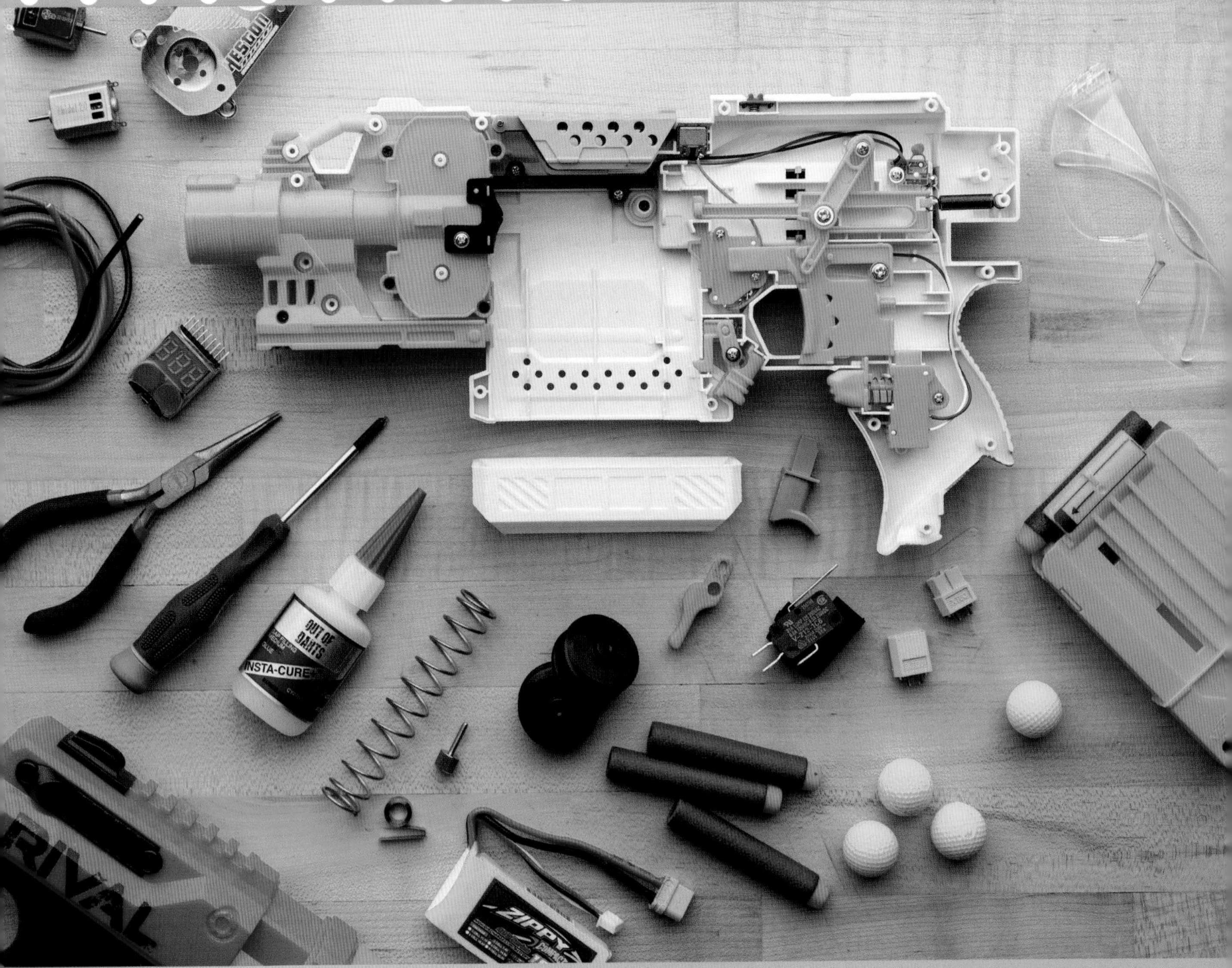
OUT OF DARTS
INSTA-CURE
RIVAL
ZIPPY

INTRODUCTION

Welcome to the World of Nerf Modding!

Nerf blasters are more popular than ever! In 2016, Hasbro sold $1.1 billion in Nerf toys. There is a large, growing community of Nerf players and **modding** fans. Also, more and more organized games are popping up across the United States and around the world. Humans vs. Zombies (known as "HvZ") is one such popular game. All players start as humans, but once tagged by a zombie, they join the horde. Played mostly on college campuses, HvZ games can last up to a week. "Endwar" attracted almost 400 HvZ players to Athens, Ohio, in 2017. "Jared's Epic Nerf Battle" in Austin, Texas, had more than 3,000 players!

I started modding rubber-band guns and building remote-control (R/C) cars when I was a kid. In 2014, I joined the online Nerf modding community. Nerf modding is a great way to learn about electronics. It also helps improve mechanical skills and develop an engineering and problem-solving mind-set. Better still, Nerf modding is relatively inexpensive and a great hobby for all ages. I play Nerf games with my young cousins, with college students, and even with people 70 and older!

Brief Modding History

People have been modding Nerf blasters for almost as long as the blasters have existed. Nerf's first dart blaster came out in 1992. It was based on the **patent** by Lonnie Johnson, a former NASA engineer and the creator of the Super Soaker.

This Sharpshooter was the first blaster I owned as a kid. It shot darts similar in size to today's Mega darts.

We have come a long way from when I was a kid. We were lucky if our Nerf blasters shot 15 feet (4.5 meters). Today's Nerf Elite and Rival blasters advertise a 90-foot (27-meter) range—and that's stock performance. "Stock" means the original blaster as sold in stores and before it's modded.

Why Nerf?

I have played paintball and **airsoft** for years. But airsoft and paintball guns are difficult and expensive to mod. They are also much more dangerous. Paintball and airsoft can't be played in parks or on school campuses. I have attended Nerf games on college campuses, at local schools, and in public parks. These large spaces and their interesting obstacles make for very fun game experiences for all ages. In addition, there are even Nerf arenas opening where you can bring your modded blasters.

Why Mod?

There are many reasons to modify a Nerf blaster. Mods can result in better performance. Mods can also make blasters look cooler, with a personal, one-of-a-kind wow factor.

Nerf blaster mods vary from a simple spring replacement to more complex rewir-

Humans vs. Zombies Endwar 2017 in Athens, Ohio. *Photo by Nick Rarri*

ing. There's also a variety of modifications to change the appearance of your blaster. I love the imagination that has shown up in Nerf modding. Paul of PDK Films says, "My favorite thing about modding is that there are no limits. You can create anything you want, and all you need to get started is an idea."

I started with a basic Nerf Stryfe mod. In this book we'll cover that very mod as well as other mod types. Get ready to dive into hands-on examples of each. The modding concepts we'll cover are for a range of skill levels and can be applied to many Nerf blasters. Those with lots of experience can skip ahead to chapters 6 and 8. I recommend that readers who are new to the hobby start at the beginning. This way you'll more easily build your modding skills.

Ready?
Let's get started!

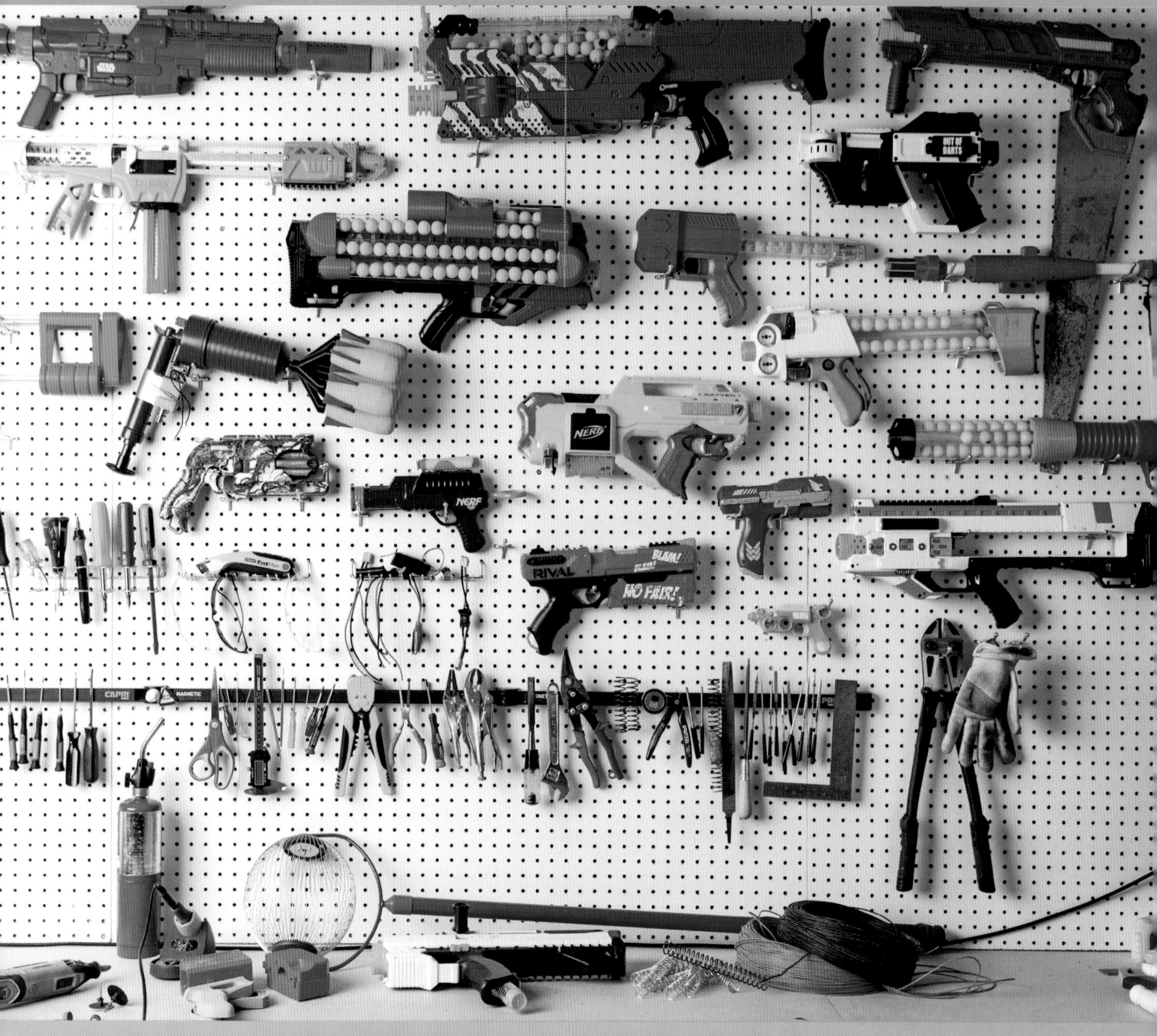

OUT OF DARTS
NERF
NERF
RIVAL
BLAM!
NO FAIR!
CAPRI
MAGNETIC

CHAPTER 1

GETTING STARTED: THE BASICS

Tools and Supplies of the Nerf Modder

You'll need a range of tools for your Nerf workshop. Each mod in this book begins with a list of required tools and supplies. On pages 12 to 17 are the main ones I use every day to mod Nerf blasters. You don't need all of these, especially when building your first toolkit.

Workshop Organization

A tidy workshop will help you complete mods faster and with fewer mistakes. Every tool in my shop has a place. Unless I'm in the middle of a project, it stays there.

Adam Savage from *MythBusters* said, "Drawers are where tools go to die." I couldn't agree more. What he meant is that tools tucked away don't get used. They are even forgotten. Tool chests hold only those tools I seldom use. Each tool has a place on my shop wall and gets put back when finished. Pliers and cutters are in one area. Knives and cutting tools have an area, screwdrivers another. I also organize small parts into bins for easy access.

I like to work on old towels to avoid scratching the blaster I'm working on. Good lighting is also important. I use inexpensive 4-foot (1.2 meter) LED shop lights.

Your perfectly tidy workspace might be a disaster area after a day of work. This happens to all of us. It's best to get in the habit of putting the workspace back to its original shape at the end of each day. Finally, I also like to keep my floors clean. That way, if I drop something, it's easy to find.

1
2
3
BOB SMITH INDUSTRIES
bsi
INSTA-CURE+
GAP FILLING
MEDIUM
GLUE
CYANOACRYLATE
2 oz
4
BLACK+DECKER
5
6
7
8
9
16 FAST SET
Clear, medium bodied solvent cement
10
AMAZING
GRIPEM
WARNING: NOT INSULATED. WILL NOT PROTECT AGAINST ELECTRICAL SHOCK
12
Sharpie
INDUSTRIAL
EXTRA FINE POINT
SUPER PERMANENT INK
Scotch
11
13
RECTORSEAL
EP-200 Epoxy Putty
Sets Hard As Steel in 15 - 20 Minutes
Plumbing • Industrial • Farm • Home • Electrical • Automotive
14
15

1. **Small flathead screwdrivers:** These are great for pulling things apart and holding parts in place.

2. **Phillips screwdrivers (#0, #1, and #2):** These are the three main sizes used in Nerf blasters. The #2 screwdrivers are for larger screwheads, such as those holding battery trays. The #1 screwdrivers are ideal for most Nerf screws, while a few smaller ones require #0. If you purchase only one screwdriver, get a #1. I like screwdrivers with a rotation cap on the handle (often called "precision drivers").

3. **Super glue:** This is great for attaching plastic parts together, but be careful not to use too much. It flows easily and can get into moving pieces, such as switches. I like the really thick stuff that tends to run less.

4. **Electric screwdriver:** This tool is a luxury, but if you mod a lot, it can save time (and your wrists). Never use a power drill to fasten screws. Drills have far too much torque (power) for the job.

5. **Snips/flush cutters:** Great for cutting wires and even bits of plastic, these can sometimes be used in place of a rotary tool or Dremel.

6. **Electrical tape:** Use this for bundling wires and in places where heat-shrink tubing can't be used.

7. **Pliers:** For removing small bits of plastic, holding on to wire connections, and pulling stock wiring terminals, pliers are your best choice. Various sizes are helpful. I love small pliers with springs.

8. **Wire strippers:** If you do a lot of modding, the self-stripping kind are very handy. If you don't have wire strippers, you can use a hobby blade to carefully strip the ends of wire as needed. This takes a little longer and more care, but it's also one less tool to buy when building your first toolkit.

9. **Acrylic cement:** This glue is incredibly strong. It actually melts plastics together. Use it carefully.

10. **GOOP:** This is a great glue because once dry it remains flexible. It's perfect for when you want two parts to have "give." Always use GOOP with gloves and in a well-ventilated area.

11. **Scissors:** Use them to cut small-gauge (thin) wires, though you will ruin the scissors over time; also great for safely opening Nerf packaging.

12. **Sharpie:** These are perfect for marking areas to be cut, drilled, sanded, or glued.

13. **Hobby knife:** Use this for cleaning up cuts, small bits of plastic, and 3D parts.

14. **Epoxy putty:** This two-part putty is kneaded (kind of rubbed) together to activate it. I prefer EP-200. Use in a well-ventilated area—it has a strong smell.

15. **Duct tape:** Nobody should be without duct tape! I use it constantly to mock up parts before gluing.

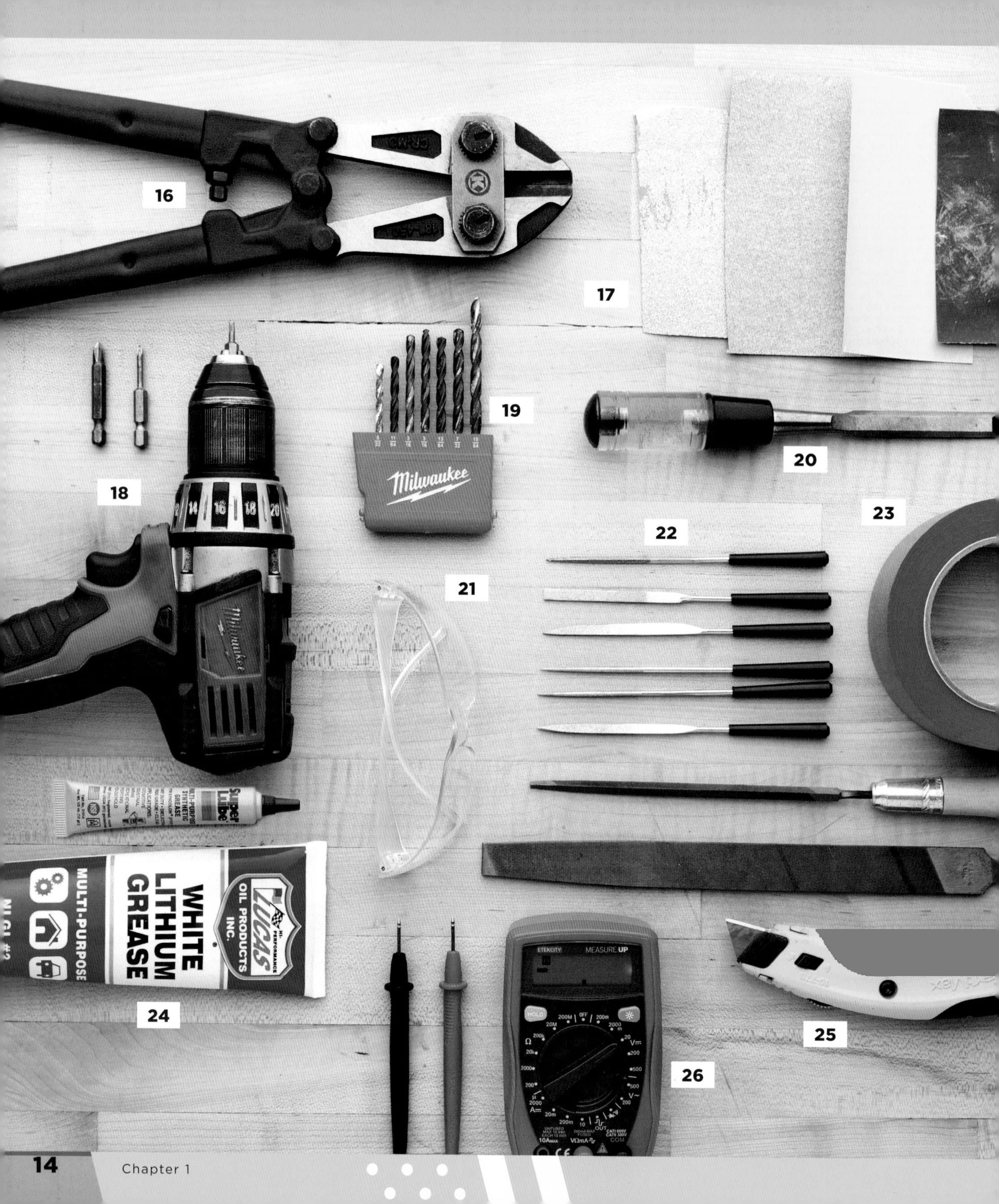
16
17
18
19
Milwaukee
20
21
22
23
Milwaukee
Super Lube
MULTI-PURPOSE SYNTHETIC GREASE
LUCAS OIL PRODUCTS INC.
WHITE LITHIUM GREASE
MULTI-PURPOSE
24
25
ETEKCITY
MEASURE UP
HOLD
26

16. **Bolt cutters:** You'll need these if you have to cut a spring. Always use eye protection when cutting anything!

17. **Sandpaper:** Most paint jobs start with some sanding. I use 100- to 800-grit for various purposes.

18. **Drill:** Making holes for wire paths is easy with a drill. A rotary tool can also do this job with the proper bit.

19. **Drill bits:** Have a variety of sizes on hand for different uses.

20. **Chisel:** An inexpensive wood chisel is great for making clean cuts in plastic.

21. **Eye protection:** Nearly every Nerf mod requires eye protection. The simple shop glasses shown are also what I use at Nerf games.

22. **Files:** Perfect for shaping parts, cleaning up cuts, and general demolition. Needle files are small and perfect for many Nerf mods.

23. **Masking tape:** Use it to cover (mask) parts you don't want painted. Also useful for holding shells together and holding other parts in place.

24. **White lithium grease:** Use this to lubricate small mechanical plastic parts. Super Lube tends to be the least messy.

25. **Box cutter:** Nerf packaging doesn't open itself! This is also good for small cuts and trimming. Use caution!

26. **Multimeter:** Use this specialized tool to check battery voltage, circuit polarity, and continuity (the continuous flow of electricity through a circuit).

27. **Soldering iron:** It's virtually impossible to complete an electronic Nerf mod without a soldering iron. This tool is used to flux (melt) the solder and connect two points. These points may include wire, switches, battery connectors, motors, and more.

28. **Wire cleaning brush:** While this removes stubborn gunk from your soldering iron, use it as little as possible—it will wear down the soldering tip quickly.

29. **Sponge:** A damp sponge is used between each solder join to remove excess solder and clean the iron. A clean kitchen sponge will work fine (just avoid scented sponges). Cut it into a few small pieces.

30. **Solder:** Solder is a metal used to join two pieces of a circuit. Use only lead-free solder! I recommend 1.5mm solder with 2-percent rosin core. (See Chapter 6 for more about solder.)

31. **Soldering iron tips:** Replacement tips are good to have on hand, though one tip should last you a while. It's a good idea to store solder and soldering supplies in an airtight container to help them last longer.

32. **Flux (cleaner):** If you solder a lot, flux paste is very handy to "tin" (clean) your soldering iron. It also wears down the tip of a soldering iron, so use it only when needed. Store it away from electrical pieces, such as circuit boards, and always close the container after use.

33. **Heat gun:** Use on heat-shrink wiring connections. (You can also use the radiant heat from a soldering iron, but it's a bit slower.)

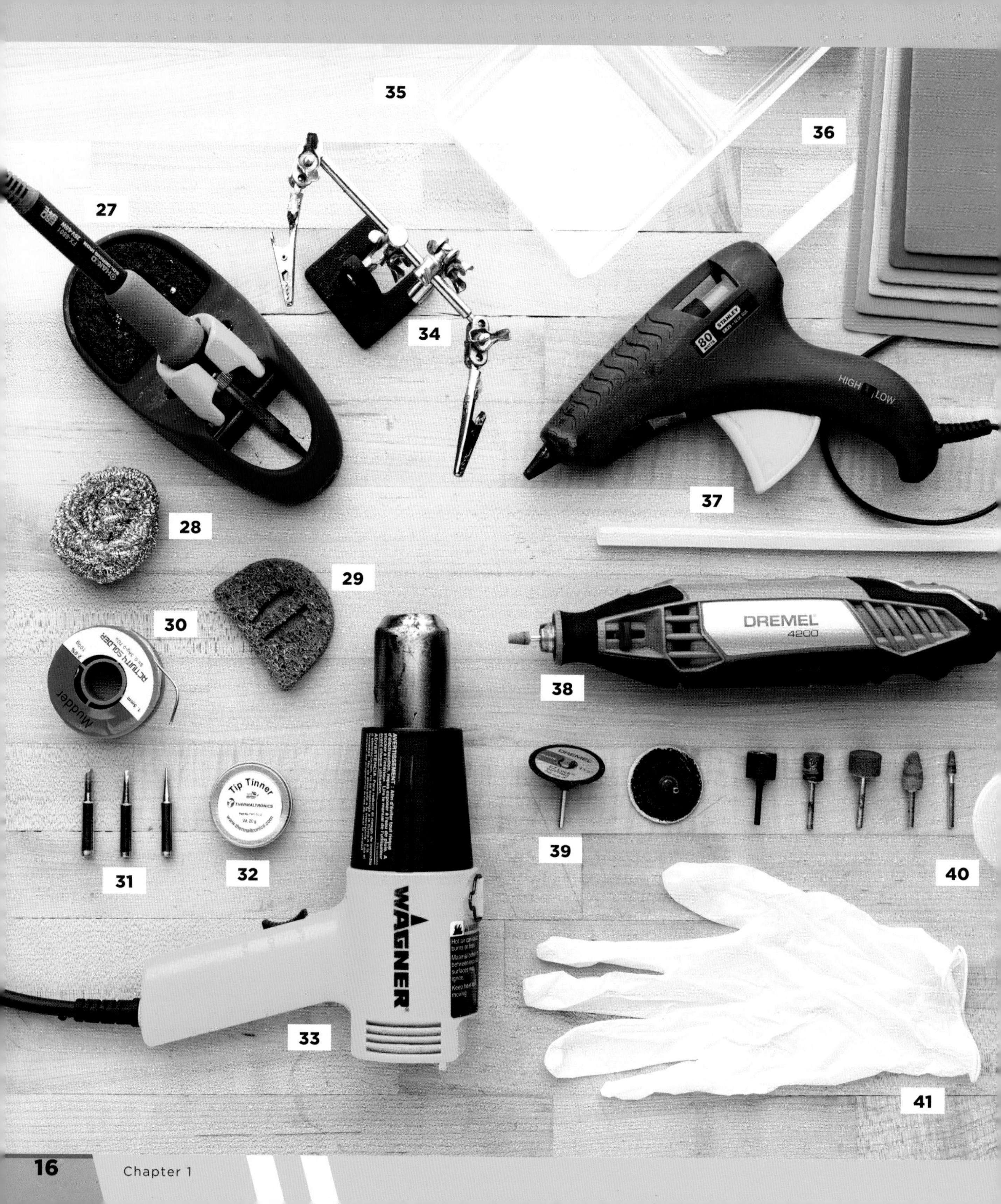
35
36
27
34
80
STANLEY
HIGH
LOW
37
28
29
30
Mudder
DREMEL
4200
38
Tip Tinner
THERMALTRONICS
39
40
31
32
WAGNER
33
41

34. **Helping hands:** These are small clamps with movable arms. Use them to hold wire, switches, and other electrical pieces while soldering. They cost as little as $5.

35. **Plastic dishes:** Organization saves time and headaches. Reuse old yogurt and food containers to organize small pieces.

36. **Craft foam:** Use it to cut out blaster mock-ups while designing. You can even use it to organize screws (just stick 'em in the foam!) for easy organization and no more misplaced screws! Also, if you kept your blaster's manual, you can stick the screws in the picture of the blaster in the corresponding spots they were removed from the real blaster. Thanks to Bobololo for this great tip!

37. **Hot-glue gun:** This is one of my all-time favorite tools. It's great for holding down wires inside a blaster. Inexpensive models are fine, but I recommend full-size guns.

38. **Rotary tool:** This is a "nice to have" tool, and once you have one you won't want to mod without it. It can be used for cutting, sanding, and grinding plastic. Dremel is the most famous brand.

39. **Rotary tool bits:** There are hundreds available, each with its own purpose. Pictured are those I use most often.

40. **Paper cups:** Perfect for mixing epoxy and other glues; Nerf YouTuber Bobololo recommends using upside-down cups as paint stands when painting small parts.

41. **Latex (or latex-free) gloves:** Great to have when kneading epoxy putty, painting, and working with other glues. Use latex-free gloves if you are allergic to latex.

I love clear pull-out bins that I can see inside without opening.

Modding Mistakes: They're Okay!

No matter how careful you are, sometimes mistakes will happen. Remember that each mistake is something you learn from for your next project. It is impossible to learn without first making mistakes.

I've made *plenty* of mistakes. Once, I wired a blaster backward. The wheels were rotating in the wrong direction. The blaster jammed and I even burned out a motor. Another time, I incorrectly wired expensive brushless motors and fried several com-

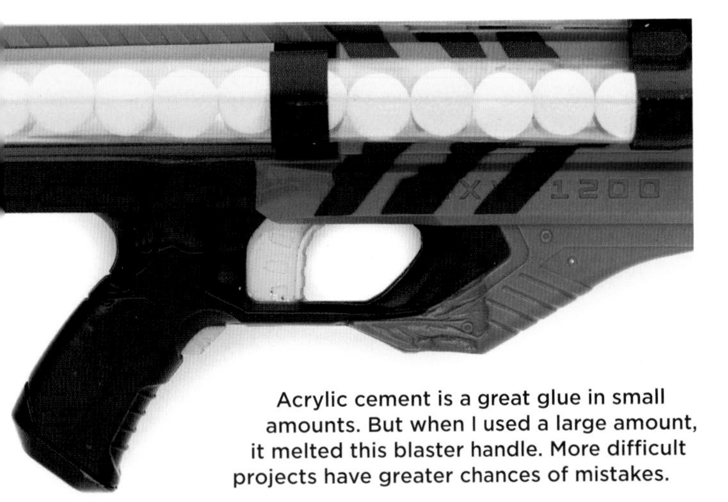

Acrylic cement is a great glue in small amounts. But when I used a large amount, it melted this blaster handle. More difficult projects have greater chances of mistakes.

ponents in a puff of white smoke. Several times I've melted plastic by using the wrong kind of glue!

If you ever find yourself getting frustrated, take a walk or grab a drink of water and a snack. Then try again later. Sometimes it's even best to sleep on it and start fresh the next day. I've been so frustrated with projects that I wanted to give up, but after a good night's sleep I finished the blasters with no issues at all!

Specs and Performance

The Nerf community spends a lot of time comparing the specifications of one blaster to another. Specifications (or "specs") are numbers used to measure a blaster's performance. Knowing how to measure performance can be helpful when deciding how to upgrade a blaster. Here are a few key specs:

fps (feet per second): This measures how far a dart will travel in one second based on the speed at which it leaves the blaster. Keep in mind, fps is affected by the dart's length, weight, and shape. Feet per second are measured by a device called a chronograph. Prices for these start at around $70. Many Nerf groups have them to share, so check with local clubs before buying one. Here are sample fps specs for some blasters:

Stock Elite	60 to 70 fps
Stock Nemesis	100 fps
Modded Stryfe	100 to 180 fps
Modded Rival	120 to 130 fps

Some homemade blasters can fire 200 fps or more. Keep in mind, fps isn't a measure of range or accuracy. It only measures how fast the dart is traveling when it leaves the blaster. Feet per second drop the farther the dart travels.

A chronograph with its fps readout.

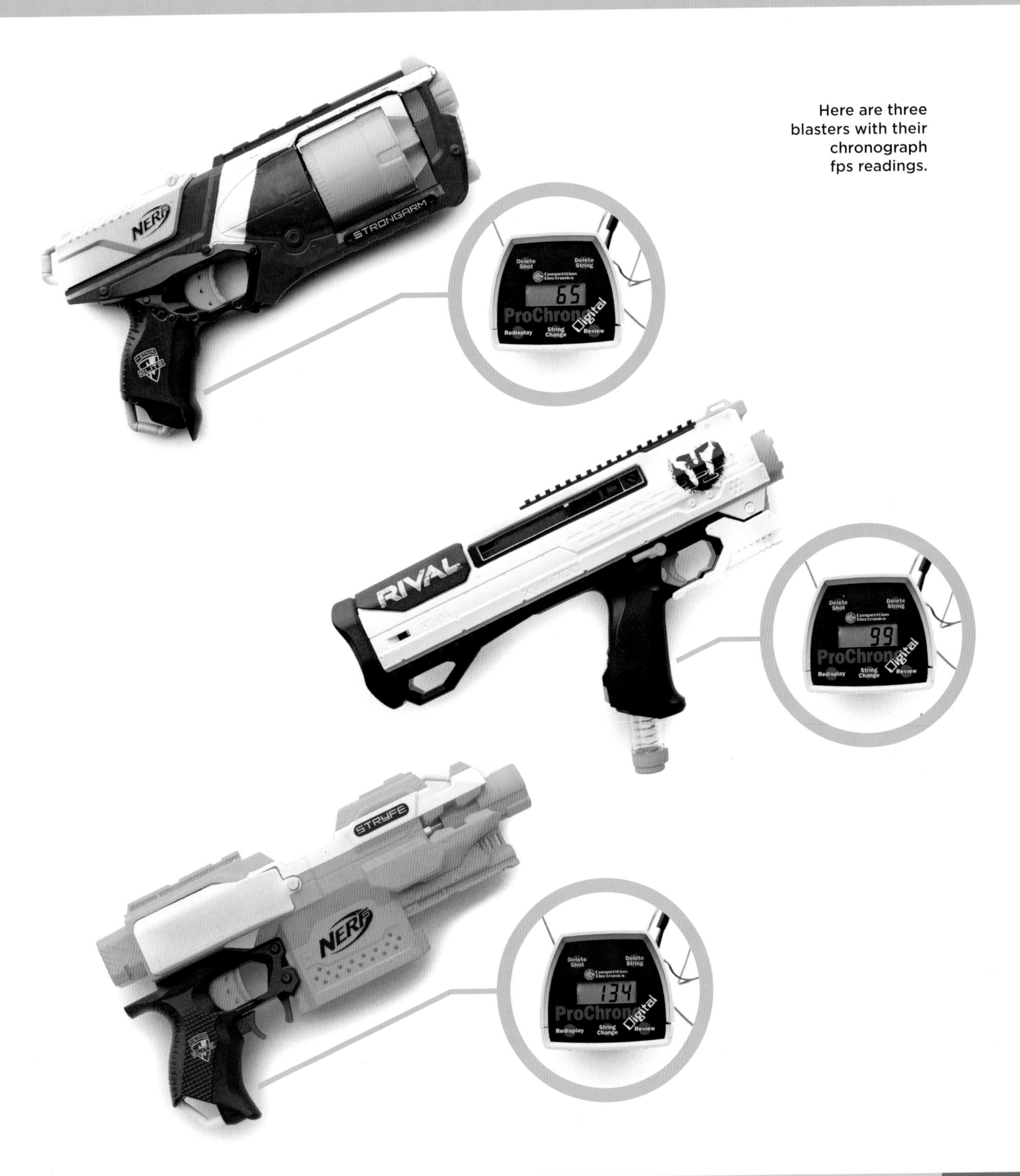

Here are three blasters with their chronograph fps readings.

Here's my workshop while I'm working on a blaster project. I created a real mess in just a few hours of work.

This is my workshop after a bit of cleanup. It is much easier to work in a clean space.

Ammo type: What type of ammo does your blaster shoot? There are many kinds. Elite darts are the most common. Rival and Mega darts are growing in popularity.

Capacity: This spec tells us how many darts, balls, or discs (also known as "rounds") a blaster can hold. A Nerf Jolt holds one dart, or round. A Zombie Strike Hammershot holds five. The Stryfe and other magazine-fed blasters hold up to 18 rounds in a **stick** magazine and 35 rounds in a **drum** magazine. I designed a back-pack blaster that can hold 1,200 rounds!

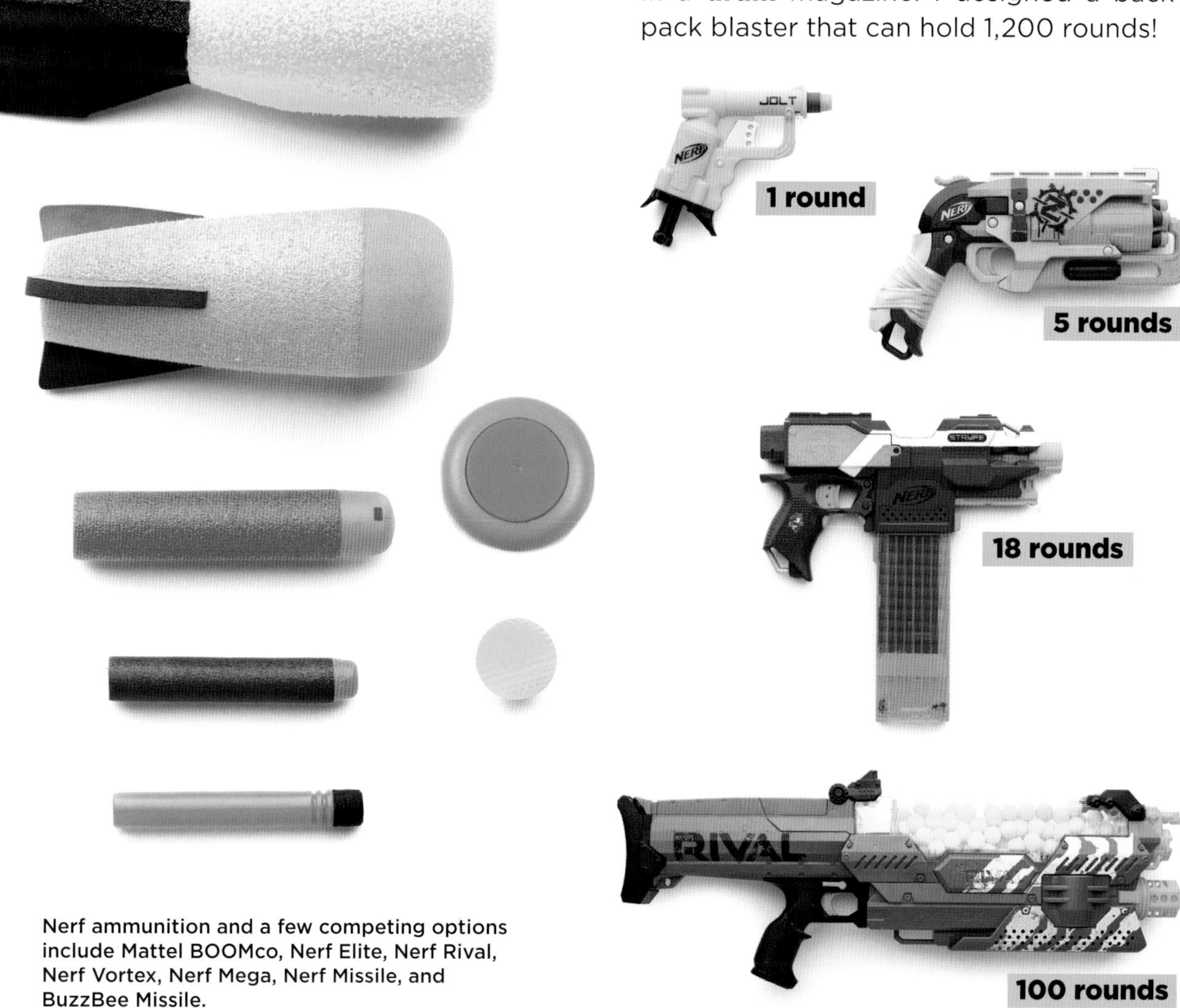

Nerf ammunition and a few competing options include Mattel BOOMco, Nerf Elite, Nerf Rival, Nerf Vortex, Nerf Mega, Nerf Missile, and BuzzBee Missile.

Range: Range is a measure of how far your blaster can shoot. My childhood blasters could barely fire across a room. Today's modified blasters can shoot 100 feet (30 meters) or more!

Sound: How loud is your blaster? Some people love the sound of a loud blaster. It's awesome during play. Others prefer quiet blasters for **stealth**. Springers, air blasters, and stringers (see Chapter 3) are the ultimate silent blasters. They don't make a sound until fired. Electronic blasters must be "revved up," or powered up, before firing. Generally, the more modified a blaster is, the louder it is.

This diagram shows user firing. The paths of two darts are shown by the dotted lines.

ROF (rate of fire): How many rounds can a blaster fire in one second? This is called the rate of fire. A stock Stryfe might be able to fire 3 rounds per second. A modded Rapid-strike can fire up to 15 rounds per second. My HIRricane Rival Zeus can fire 20 rounds per second!

Accuracy: This is just as important as how fast and how far your blaster shoots. Rival blasters tend to be more accurate than stock dart blasters. Hasbro Elite AccuStrike darts have proven more accurate than older darts. Mattel BOOMco darts are the most accurate dart I've ever shot. Short darts are very accurate too, but can only be fired from specific blasters.

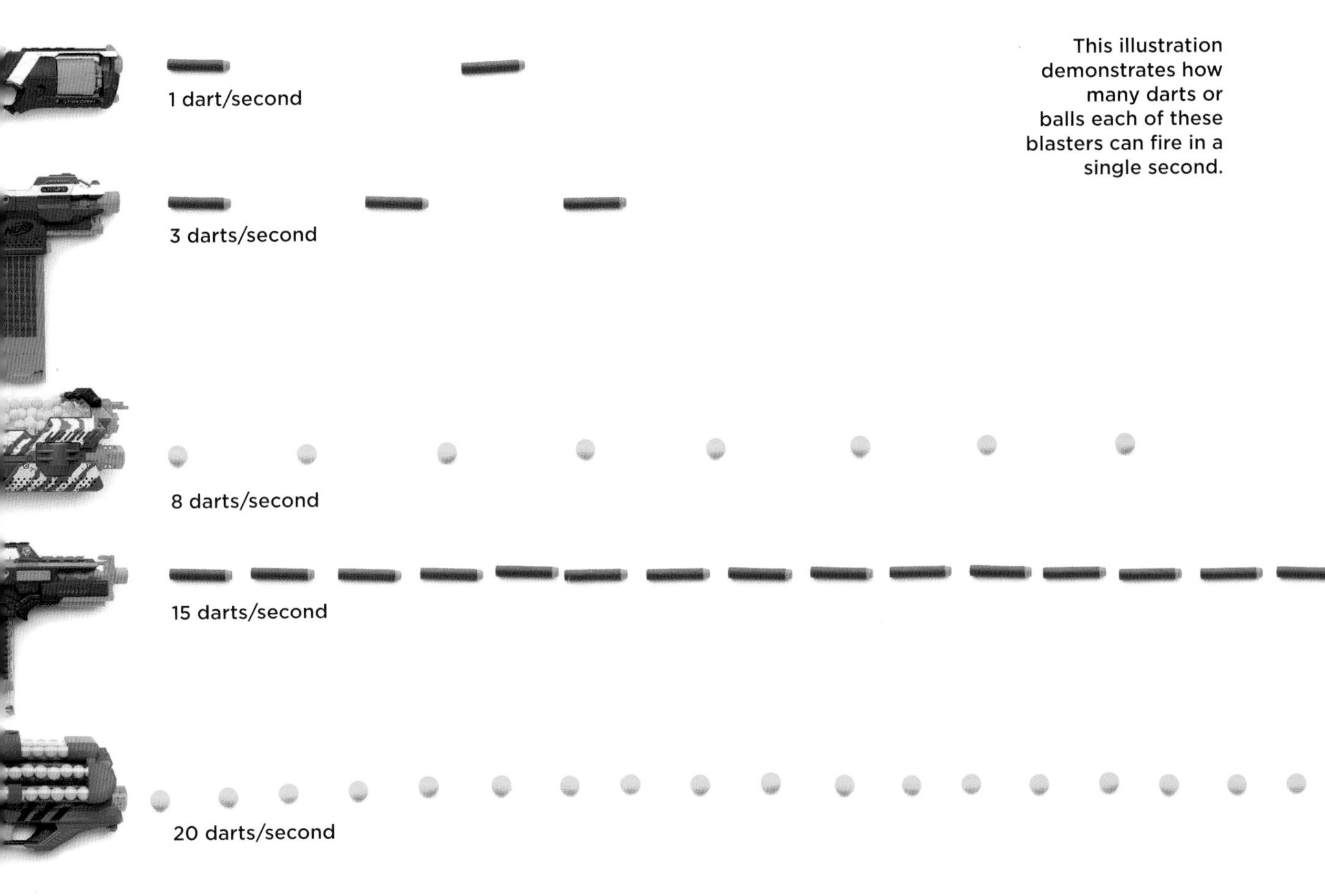

This illustration demonstrates how many darts or balls each of these blasters can fire in a single second.

Soldering

Soldering irons are extremely hot (600–900°F, or 315–455°C). Protective eyewear is important. Hot solder can easily flick toward your face and eyes.

Soldering irons should be kept on a clean surface with the cord unobstructed. And don't forget to unplug them when you're done! I have burned myself just twice in the last two years after soldering more than a thousand joints. Practice makes perfect. Being aware of safety from the start is a big help. Use only lead-free solder. Lead can cause all kinds of health issues.

Sharp Tools

My worst injuries have come from using hobby knives and 3D printer removal tools. First, always cut away from yourself. Sometimes you may think you're cutting away from yourself, but you have a finger holding a part that's in the path of the blade. Also, put away sharp tools that aren't in use, and use their protective coverings. Finally, always wear eye protection when cutting, grinding, sanding, or using a rotary tool. Comfortable protective eyeglasses cost as little as $2 a pair.

It's important to keep a soldering iron and its cord free of obstructions and safely stored.

When using sharp tools, always cut away from the body. Also be mindful of where your fingers are in relation to the path of sharp blades. Here I have repositioned my fingers to be out of the cutting path.

These glasses are very inexpensive and available at my hobby shop, at www.amazon.com, and at local hardware stores.

During gameplay—
especially in public spaces—
I encourage players to use
the term **blaster** instead
of **gun**. I also suggest the
phrase "I **tagged** you"
instead of "I **shot** you."

Gameplay

All sports have safety equipment, and professionals always use it. Darts are only so accurate, so it's not always possible to avoid eyes and faces. I wear standard shop safety glasses to every single Nerf game I play in. They are affordable and tough. I also require them at any games I host.

Painting

Paint only in a well-ventilated space. Wear a respirator or mask to avoid inhaling paint fumes. I recommend painting outdoors when possible. Read all directions and warnings on paint cans. And do not use lead paint. Lead paint is rare these days, but it is a good warning to remember.

Electricity

Always make sure your hands are dry when working with electrical circuits. Water provides a conductive path for electrical current. DC (direct current) in Nerf blasters isn't enough to electrocute you, but it can give you a shock or even a burn.

Also, household outlets run on AC (alternating current) and easily have enough current to electrocute you. Never plug anything into an outlet other than an appliance with the correct plug!

LiPo Batteries (Lithium Polymer/Ion)

LiPo batteries are safe if you follow a few guidelines. First, LiPo batteries should be handled with extreme care. They are dangerous when dropped, cut, burned, or otherwise damaged. If the battery's individual cells are exposed to air, it causes a violent chemical reaction. Careful handling also increases their life, allowing you to get more bang for your buck!

LiPo batteries are the key to high-end electric Nerf mods. Smartphones and other devices have control circuits that tell the device to cut off power if the battery is draining or overheating. (Maybe you remember the Samsung Galaxy Note 7 smartphone fires of 2016.) Our hobby-grade LiPo batteries are even more powerful than phone batteries. They also lack control circuits that prevent overheating. It is important to monitor the voltage of your batteries.

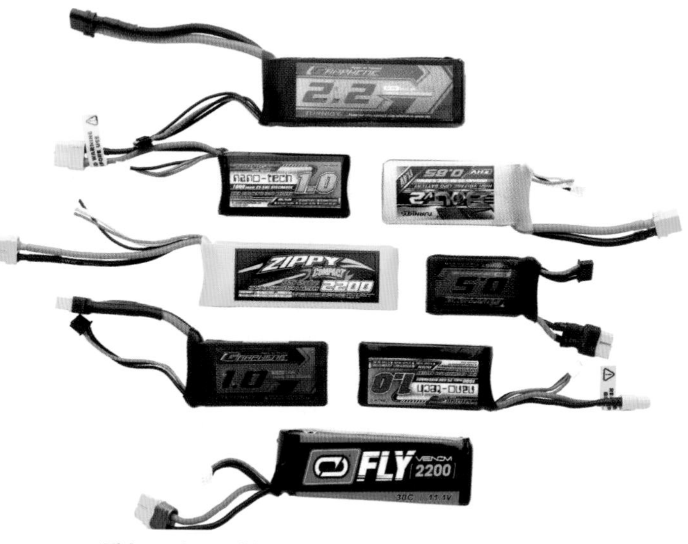

This variety of batteries is used in many of my Nerf blasters. See Chapter 7 for more information.

HERE ARE A FEW OTHER LiPo RULES:

Do not overdrain: If a battery drains below 3 volts, it might not take a charge again. Voltage alarms make a very loud buzzing noise when any cell drops below 3.3 volts, telling you to charge the battery.

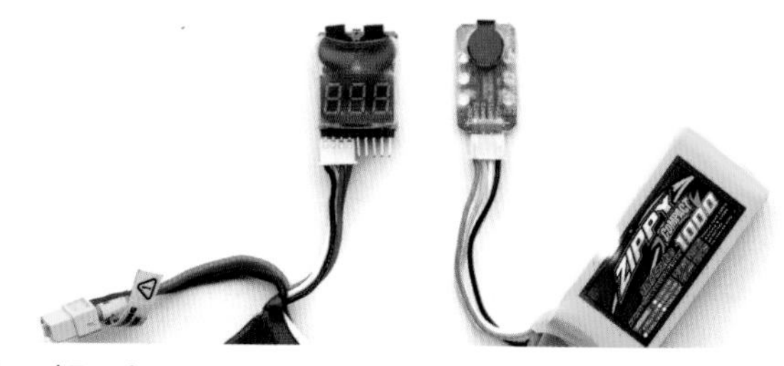

Two $3 voltage alarms. One shows the voltage readout of each battery cell. The other just has an audible alarm. Either one works well for our purposes, although the smaller one fits more easily in some blasters.

Charge in a fireproof container: LiPo batteries should always be charged in a LiPo safety bag or other fireproof container. I have never seen a charging battery catch fire. However, I once over-drained a battery by accident, ruining the pack. Charging batteries should never be left unattended. Charging containers should be breathable (no airtight containers). Charge away from flammable materials. I charge my batteries in the middle of my garage floor, away from the walls.

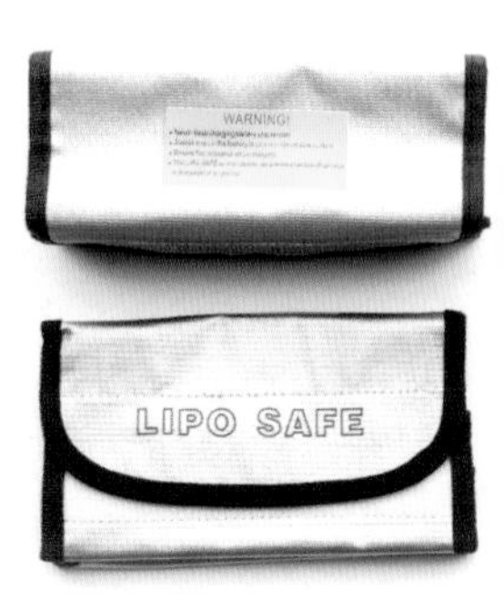

Do not short out the battery: LiPo batteries can make extremely high electrical current (see more about electricity in Chapter 6). Never let the battery's positive and negative leads touch. This is called "shorting out" the battery. Shorting can cause sparks, burns, and even fires. It is also the fastest way to ruin a battery. No matter how experienced you are, I recommend testing new wiring jobs with a low-voltage, low-current battery harness, such as a AA or D cell alkaline test circuit. This will prevent you from shorting a battery due to a wiring mistake.

Shipping charge: Batteries are shipped with a 30 percent charge. This is for safety reasons. A new battery should not be used until properly charged.

Storage charge: Batteries should be discharged to "storage charge" (approximately 3.7 volts) after use. This is safer and it makes batteries last much longer. If your battery charger doesn't offer a storage charge, simply run down your battery during gameplay. I often find that after a day of play the battery voltage has dropped to storage charge or less simply through use. Also, don't charge extra batteries you won't need.

Balance charging: This is the practice of using a charger that can charge each cell to a specific voltage. This is safer and better for the batteries than charging the whole pack to a combined voltage. This charging is done using the balance port, which is the smaller plug on your LiPo battery. Read more about LiPo batteries on page 62.

Disposal: LiPo batteries should not simply be thrown into the trash. Contact your trash service for information on how and where to properly throw out batteries.

Puffy or damaged battery: Never use a puffy battery pack. This usually happens from overcharging, from physical damage, or by pulling too much current from a battery. Damaged packs need to be disposed of properly.

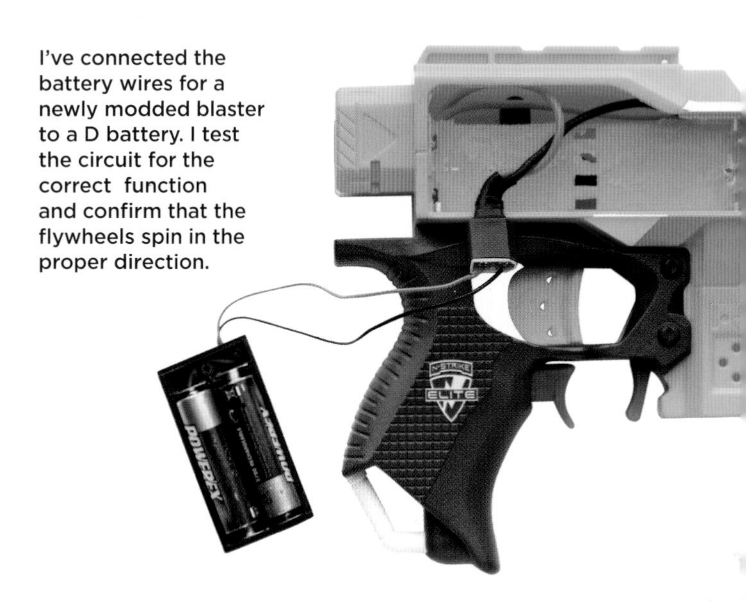

I've connected the battery wires for a newly modded blaster to a D battery. I test the circuit for the correct function and confirm that the flywheels spin in the proper direction.

Temperature: Store LiPo batteries at room temperature. Don't leave them in hot cars or near heat sources. LiPo batteries do not work well in cold temperatures. It's not dangerous to use a LiPo in the cold, but it will cause shorter run times and lower performance. In the winter, it may be so cold that your blaster will not fire. Do not charge your battery in a cold environment.

Airplanes: Bring LiPo batteries in carry-on luggage but only at storage charge and in a LiPo bag. (Do *not* carry on your Nerf blaster—it will be confiscated or refused.)

Springer blasters

CHAPTER 3

TYPES OF BLASTERS AND HOW THEY WORK

Nerf uses four primary sources of **propulsion** in current blasters: springs, electricity, air, and elastic. Each has its own advantages in gameplay. Each also has its own challenges when it comes to modding. For example, spring-powered blasters (called "springers") require no electrical knowledge and can be very simple to mod. Electric blasters have the highest fire rates but require soldering and basic electrical skills.

In this chapter, we will look at all four types. Then, in chapters 4 and 7, we'll cover springers and electric flywheel blasters in detail.

PRO ADVICE

Nerf YouTuber Jangular says, "Opening up a blaster for the first time can be daunting. But we all make mistakes that help us learn. Don't let fear of failure prevent you from trying something new. Dive in and have fun!"

Springers

The springer is the most common kind of blaster. There are more springers than all other blasters combined. It is also the oldest type of dart blaster. The first was the Sharpshooter, which came out in 1992.

The design is simple. A spring and a plunger tube compress air, which forces the dart out of the barrel. Think of this like blowing a wrapper off the end of a drinking straw.

The user generally has to "prime" a springer-style blaster before firing. This is done by compressing the spring. The trigger then releases this stored energy. A trigger catch holds the spring in tension until the blaster is fired. Some springers don't have a trigger. They feature an automatic release once the blaster is fully primed. These blasters are less popular due to the difficulty in aiming while priming the blaster at the same time.

It's worth noting the two plunger types used by Nerf:

- A **direct plunger** is lined up with the firing chamber.
- A **reverse plunger** has an entire plunger housing that moves when fired. These are far less efficient and are uncommon in current blasters. The upgrade process remains the same, however.

Here are a few springs, all from Nerf blasters. Notice that their sizes and shapes vary widely. Each has a different purpose.

SPRING-POWERED NERF BLASTERS

- Maverick
- Strongarm
- Zombie Strike Hammershot
- Elite Alpha Trooper
- Recon MKII
- Longshot
- Rival Helios
- Rival Apollo
- Rival Atlas
- Rival Artemis
- Rival Kronos . . .

. . . and hundreds more!

Electric Flywheel Blasters

Electric Nerf blasters are usually called flywheel blasters. These have spinning wheels inside (the flywheels), usually powered by electric motors. Flywheels are used in baseball pitching machines. If you look closely, you'll also see them used to move balls at the bowling alley.

Nerf flywheel blasters have two wheels that spin in opposite directions. The dart is caught between the two wheels and propelled out of the blaster. Each flywheel has its own motor connected to the same

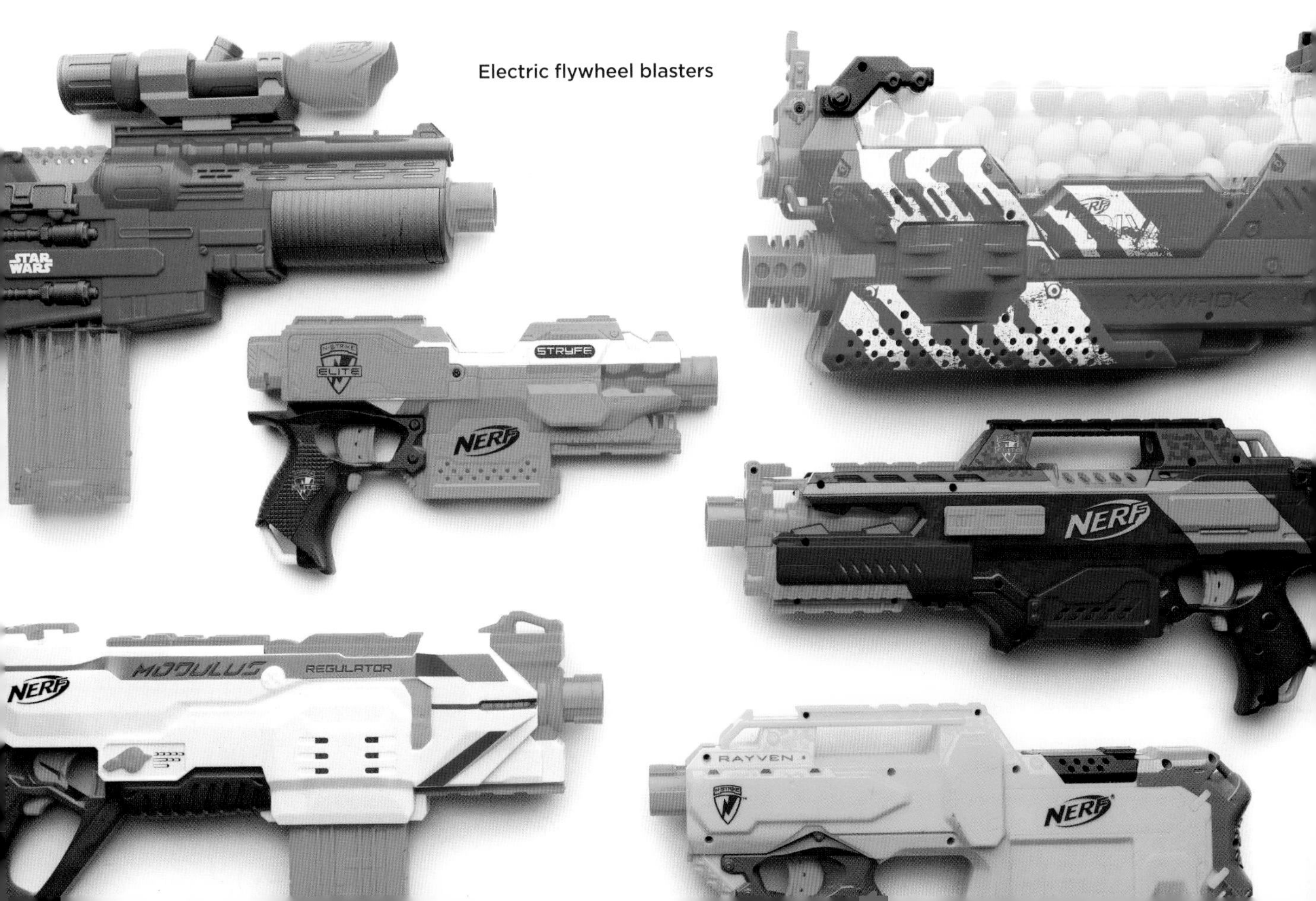

Electric flywheel blasters

electrical circuit. Flywheel blasters almost always use a battery, but a few are hand-powered, such as the RevReaper and the discontinued Ripsaw and Buzzsaw blasters.

Electric blasters need a way to guide a dart into the spinning flywheels. There are two methods: a manual pusher and an electronic pusher. Manual pushers, as in the Nerf Stryfe (pictured), use a lever and your trigger finger to fire.

Other blasters, such as the Rapidstrike, use an electronic pusher. Here, a motor turns, pushing a gear in and out. You may have seen a similar type of motion on steam-powered locomotives.

Belts are another example of electronic pushers. A belt turns in one direction and guides each dart or ball into the flywheels.

Here are a few of the dozens of flywheel cages and flywheels available. Flywheel cages hold spinning flywheels, which propel darts in electronic blasters. The orange one in the center is a stock Nerf Stryfe cage. All the surrounding ones are available mods. Clockwise from top left: Open Flywheel Project (OFP) Rayven cage with Worker flywheels; Black Steel Props cage with Cyclone flywheels; Artifact cage with Artifact aluminum flywheels; OFP Rayven cage with Blasterparts flywheels; Heston Systems Engineering CNC cage with Heston flywheels; and OFP Stryfe cage with Out of Darts Insutanto flywheels.

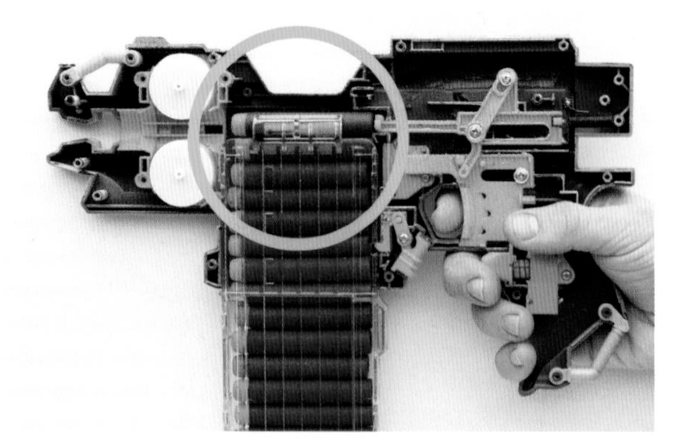

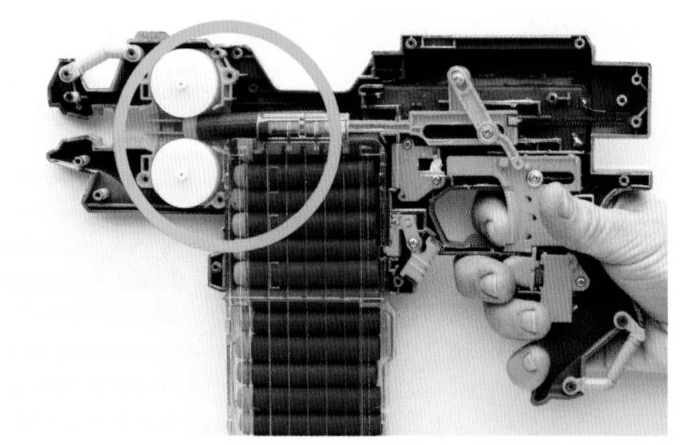

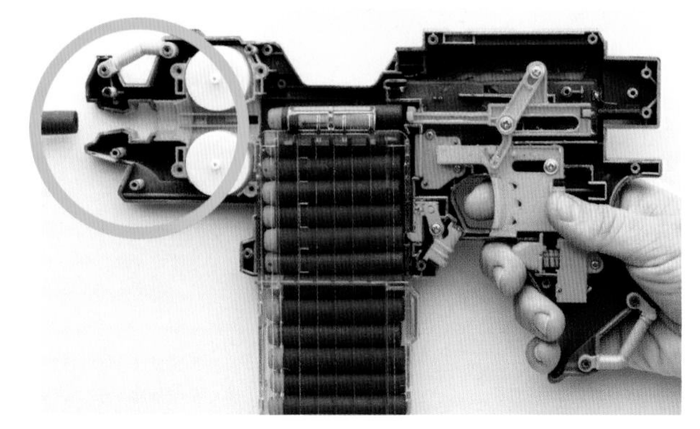

Here you can see a dart passing through spinning flywheels. As you pull the trigger, a lever connected to the trigger moves the pusher forward, which pushes the dart into the flywheels. As the dart is pinched between the flywheels, they transfer energy to the dart, propelling it out of the blaster.

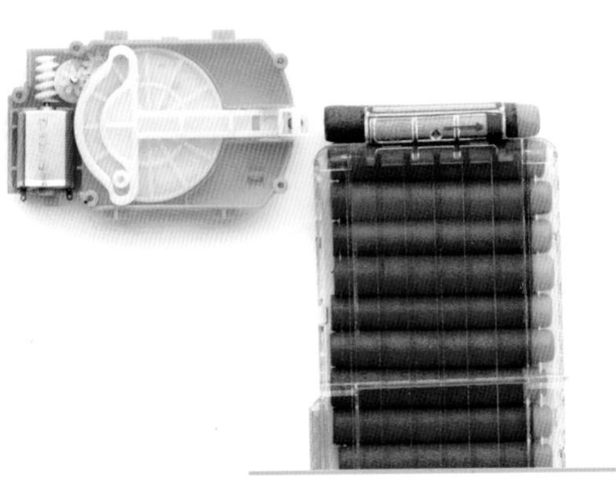

This gearbox is from a Rapidstrike. As the motor turns, the "slip eccentric" gear pushes in and out. This is similar to some steam locomotive trains.

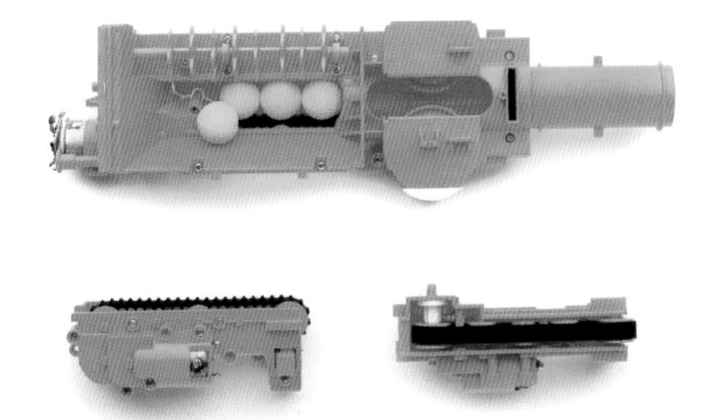

Here are two examples of Nerf belt-fed pushers: a Rival Nemesis and a Rival Khaos. Both are full auto.

FLYWHEEL-POWERED NERF BLASTERS

- Styfe
- Rapidstrike
- Infinus
- HyperFire
- Rayven
- Rayvenfire
- Hailfire
- Barricade
- Mega Mastodon
- Rival Zeus
- Rival Khaos
- Rival Nemesis
- Rival Hera
- Rival Prometheus

The discontinued Nerf Stampede has both a battery and a spring. Its electric motor powers a spring-and-plunger system. It has a really fun-sounding firing action.

Air-powered blasters

Air-Powered Blasters

Air-powered Nerf blasters are more uncommon, but several are still available. Air-powered blasters store air pressure in a tank. A **valve** connected to the trigger releases air pressure, transferring the energy to the dart.

Air blasters are extremely popular to modify. They have a lot of power. They can also be simply modified to fire a wide variety of projectiles.

Some Nerf modders add air tanks to other kinds of blasters. For example, the main blaster might be powered by a flywheel, but it also has a rocket launcher powered by an air tank. In gameplay, these are used as "shield busters." This means they are the only way to destroy a shield in gameplay. This makes for a fun and challenging game!

We're not going to learn how to mod air-powered blasters in this book. They just aren't as easy to find as electric-and spring-powered blasters. Very few new air-powered blasters are available for sale. However, to give you an idea, one popular air blaster mod is called re-barreling. This is when we modify a blaster that fires one ammo type and to fire another ammo type. For example, if a blaster meant to fire a rocket (very large and heavy) is modified to fire a dart (much smaller and lighter), the dart will travel a very long distance at a high fps. This is because the blaster transfers the same amount of energy used to fire the rocket to a much smaller and lighter dart.

AIR-POWERED BLASTERS

- Nerf Mediator barrel extension
- Nerf Magstrike
- Nerf Titan AS-V.1
- Nerf Wildfire
- BuzzBee Big Blast
- BuzzBee Extreme Blastzooka (XBZ)

This blaster, modded by Captain Xavier, has had an air tank added as a secondary rocket launcher. It fires Nerf Demolisher Missiles.

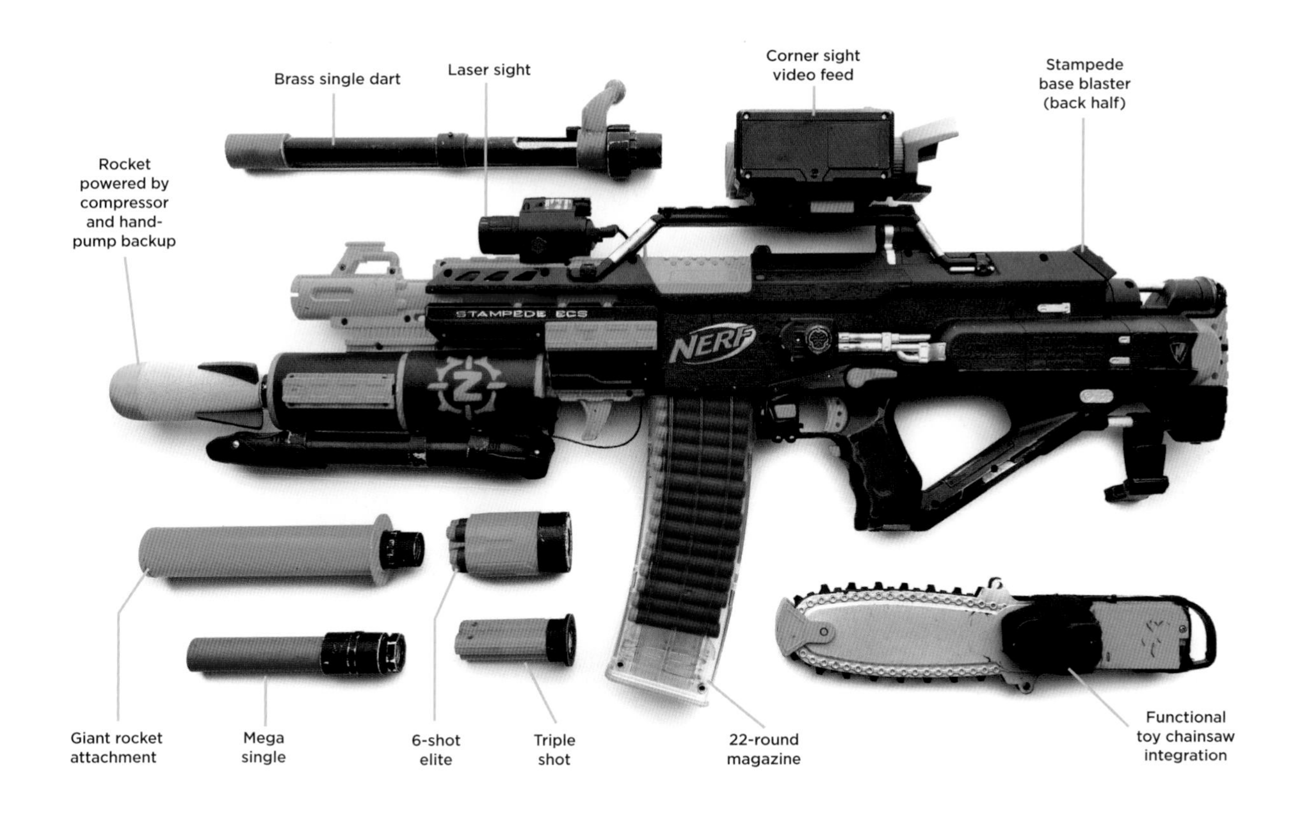

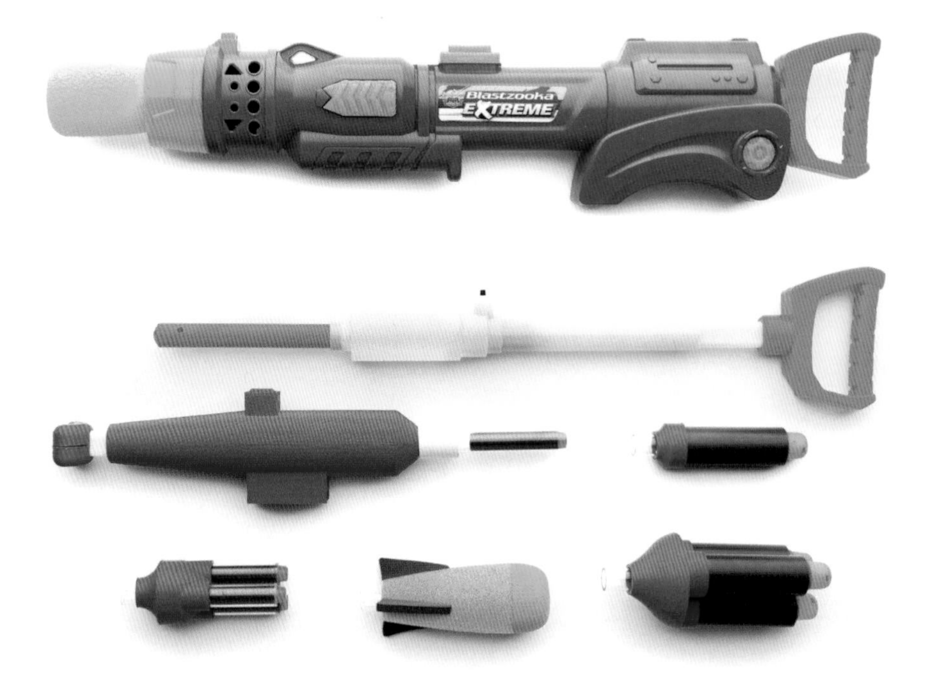

The blue blaster is an Extreme Blastzooka (XBZ). Below it is the white air tank taken from inside. The purple blaster is a custom-made, 3D-printed minimization designed by my friend Tarik. Minimization is the process of reducing the size or shape of a blaster. You can see how much smaller this purple blaster is compared to the original. It has attachments for single Elite, Mega, Triple Elite, Triple Mega, and Demolisher rockets.

String-Powered Blasters

Some blasters use elastic string to fire darts. Elastic works like a spring, but a true elastic-powered blaster can be very quiet when firing.

Some blasters that look like they are string powered, such as the Zombie Strike Crossfire bow and the Big Bad Bow, actually have springs and plunger tubes. The string is there for looks.

We won't cover elastic-powered mods, specifically, but they are easy to modify. Usually this is done by adding elastic, rubber bands, or rubber tubing to the existing elastic. These materials are all available at local craft and hardware stores.

This Big Bad Bow blaster actually has a spring and plunger inside. It's not really a string-powered blaster. You could remove the strings and "bow," and the function would be unchanged.

String-powered blasters

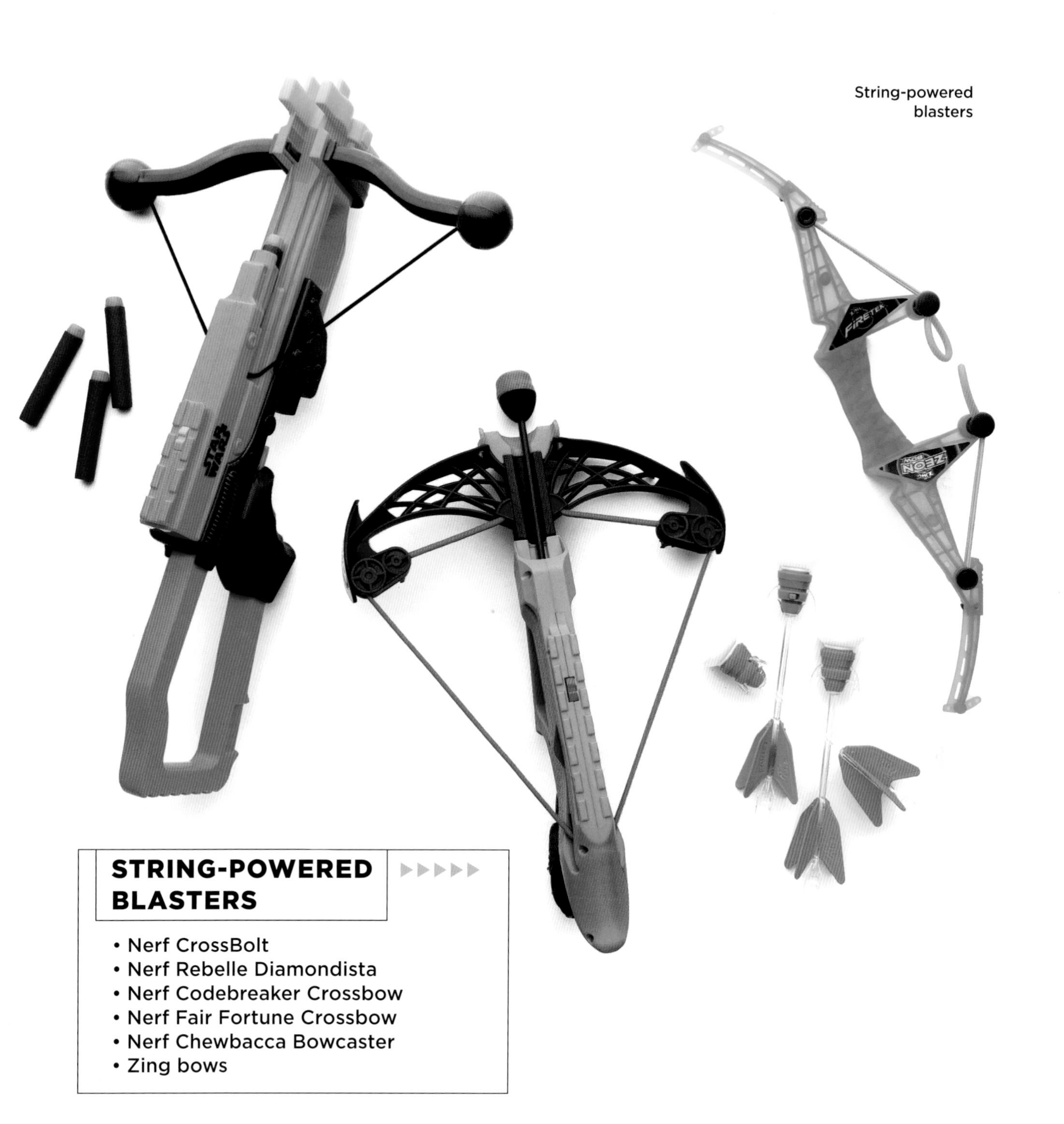

STRING-POWERED BLASTERS

- Nerf CrossBolt
- Nerf Rebelle Diamondista
- Nerf Codebreaker Crossbow
- Nerf Fair Fortune Crossbow
- Nerf Chewbacca Bowcaster
- Zing bows

This is a standard compression spring. This one is commonly called a K26. They are widely used in many Nerf blaster upgrades. At least a hundred blasters can use this spring in varying lengths.

This is a typical extension spring.

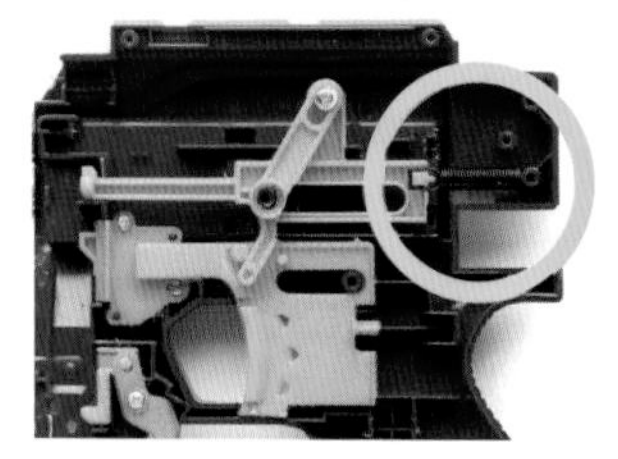

Here you can see an extension spring in a Nerf Stryfe. This spring returns the trigger and mechanical pusher back into position after each shot.

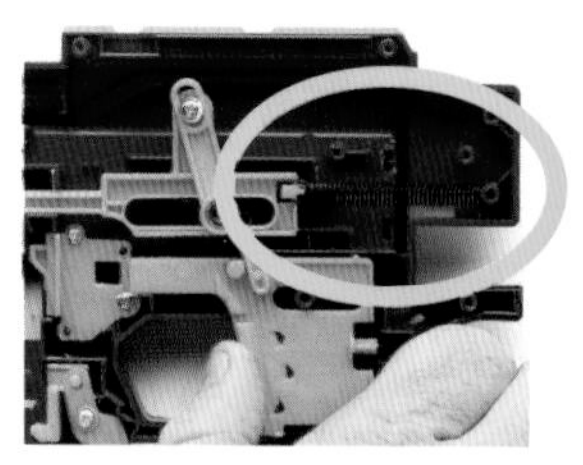

When the trigger is pulled, the extension spring provides the return force for the trigger assembly's simple lever and slide.

This is a torsion spring from a Nerf Rival Zeus. It pushes a flap in the firing system that prevents a second ball from entering the flywheels.

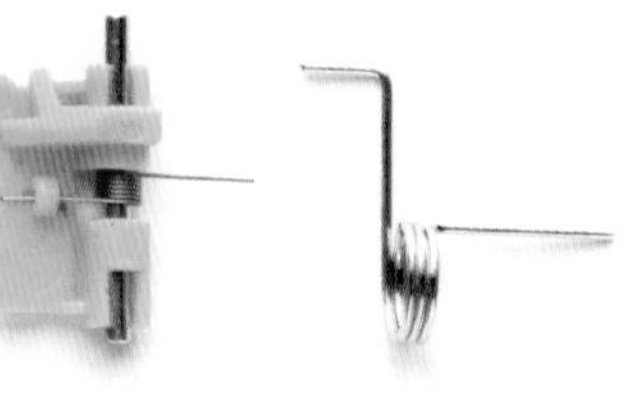

This constant-force spring upgrade is made by Foam Blast. It is from the drum magazines pictured.

These drum magazines have this tiny constant-force spring inside. It feeds the darts through the magazine.

Compression: A compression spring is squeezed, or "compressed," to store energy. Compression springs are used in almost every spring-powered blaster. We will cover compression spring mods step by step in the next chapter.

Extension: Extension springs are the opposite of compression springs. They are stretched, or "extended," to store energy. Extension springs are used in many blaster firing systems. These springs start with the coils touching each other and are then stretched or extended.

Torsion: Torsion springs store rotational energy by twisting. They are often used for levers and other parts of a firing system. Because torsion springs rarely affect a blaster's performance, they are seldom modified.

Constant force: Constant-force springs often look like a ribbon. When uncoiled, they store energy because the spring wants to return to its coiled state. Constant-force springs are common in vintage windup toys and music boxes. They are used in Nerf drum magazines to move the darts around the drum's circular dart path.

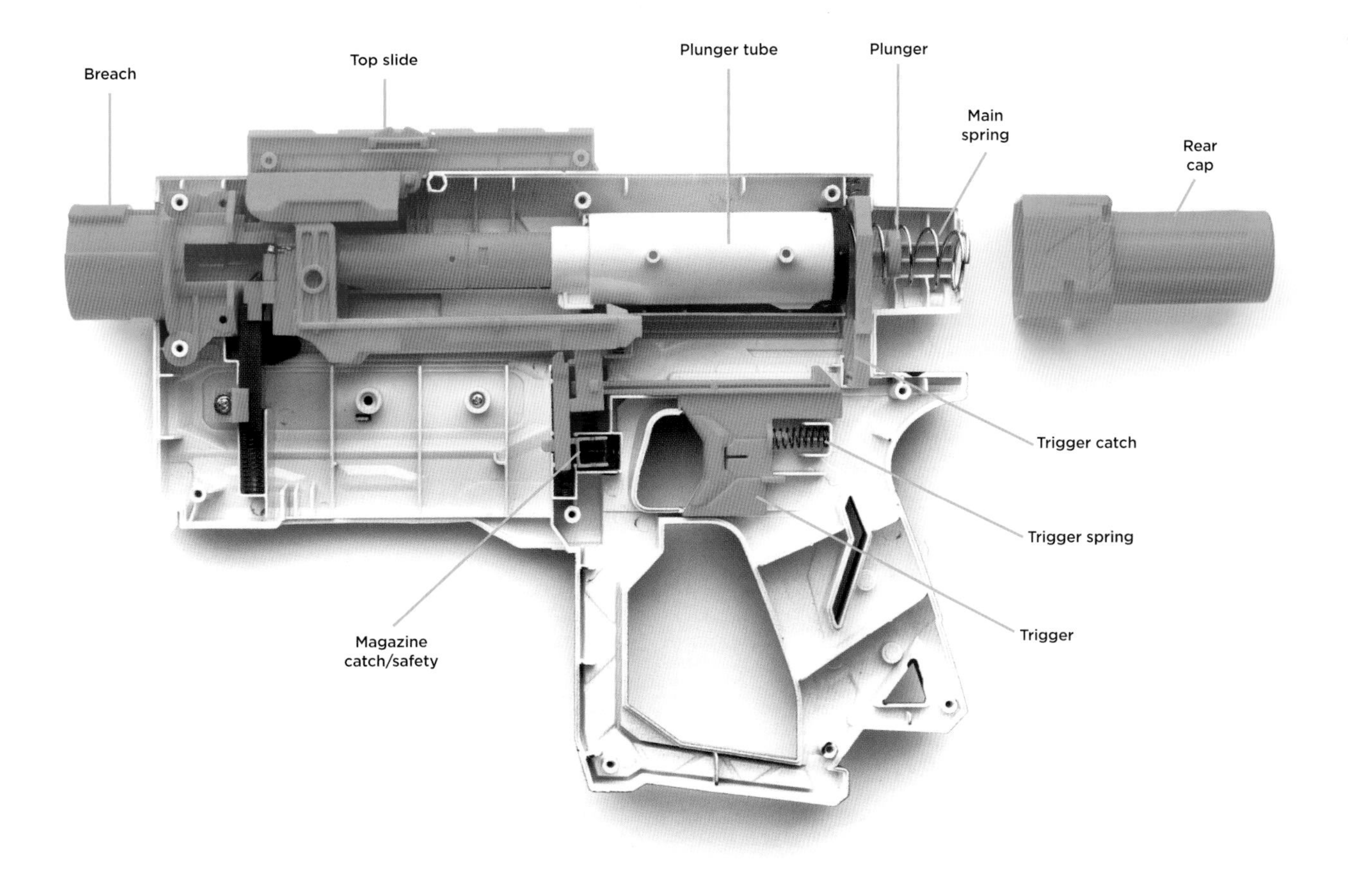

Springer Anatomy

The photo in this section shows the inside of a Nerf Recon MKII. This is a common springer layout. The trigger connects to a catch. The catch is the release mechanism for the plunger/spring assembly. Generally, this catch has some sort of lever and mounting point. Sometimes the catch and its spring must be reinforced to support higher-powered springs.

Replacing springs is fairly simple. First, we open the blaster, being careful to keep track of where all the parts came from. Then we remove the stock spring. If we want, we can lubricate the plunger tube with Super Lube or white lithium grease. Then we simply replace the spring and reassemble the blaster. Some of these mods can take as little as 5 to 10 minutes. Others can take an hour or more, depending on the blaster.

The Nerf Recon MKII is a popular springer.

A wide range of ready-to-use Nerf upgrade springs are available online. They often come with parts to upgrade the trigger and catch. They can also take a lot of the guesswork out of modifying a spring-powered blaster.

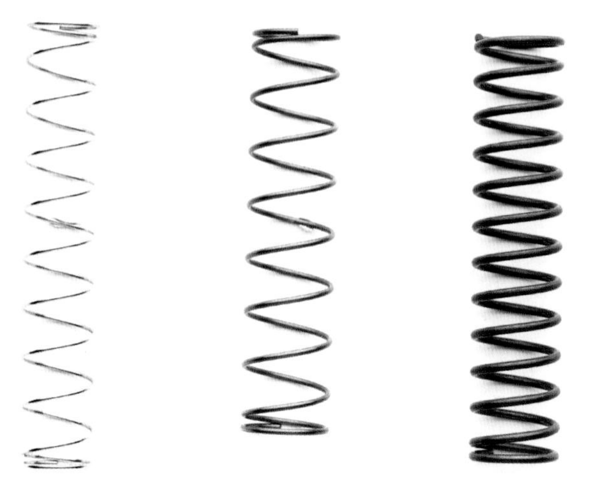

These springs are similar size but have different thicknesses and weights. They are rated at about 3, 7, and 9 kg.

Spring Ratings

We rate springs by the amount of force required to compress them a certain distance. This is usually measured in kilograms (kg). A stock Hammershot, for example, has a stock spring rated at about 3 kg and can take an upgrade spring rated at 7 kg. A number of manufacturers rate springs differently, so use this as a reference point only. Local hardware stores carry some springs, but you will find a better variety at online Nerf shops or on Amazon. For more resources, check out "Favorite Shops" on page 115.

Concerns

Any time we upgrade a blaster, there is a risk of damaging the blaster if we take the mod too far. An extremely strong spring can potentially break the blaster's internal plastic parts. This is why we often have to reinforce the plastic. Sometimes we also need to replace the trigger catch spring so it is strong enough to hold the upgraded spring. Kits sold online have usually been tested and are safe for your blaster.

PRO ADVICE

My friend Captain Xavier, a popular YouTuber and the King of Springs, has great advice: "Don't get greedy. Over-modding your blaster can result in it breaking and leaving you with nothing." What he means is, don't simply shove the largest spring possible into your blaster.

Time: 10–15 minutes

NERF TRIAD

TOOLS AND MATERIALS

- #0 screwdriver
- Triad upgrade spring
- White lithium grease or Super Lube (optional)

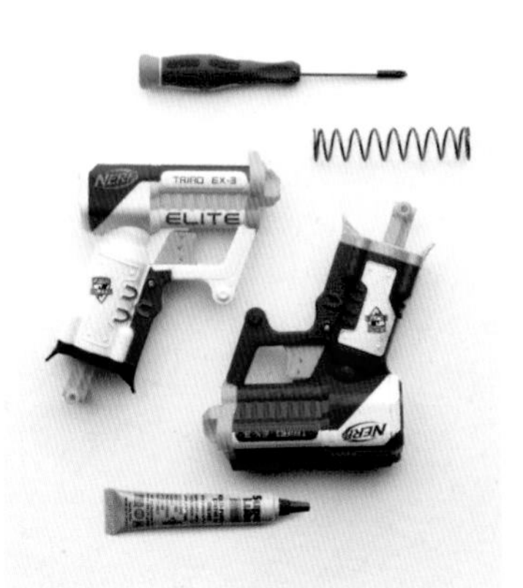

The Triad is a surprisingly capable blaster. I once lasted four hours of Humans vs. Zombie play as a human with just a Triad after my main blaster jammed. It's quick to reload and has a three-shot capacity.

First, open the blaster: use the screwdriver to remove the four screws at the blaster base.

The Nerf Jolt and Triad are the smallest of all blasters. They are also probably the easiest spring upgrade ever. This mod covers the Triad, but Nerf Jolt and MicroShot blaster modifications are nearly identical. All these blasters can be upgraded with a premade upgrade spring. Or you can use two stock springs.

The MicroShots series blasters are Jolt re-skins. A "re-skin" is when the same inside parts are put into a newly designed shell. These are often marketed as different blasters.

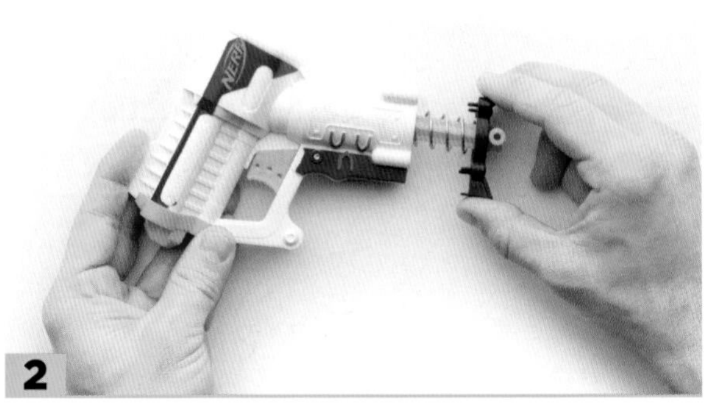

When all the screws are removed, pull out the plunger tube assembly and set it aside.

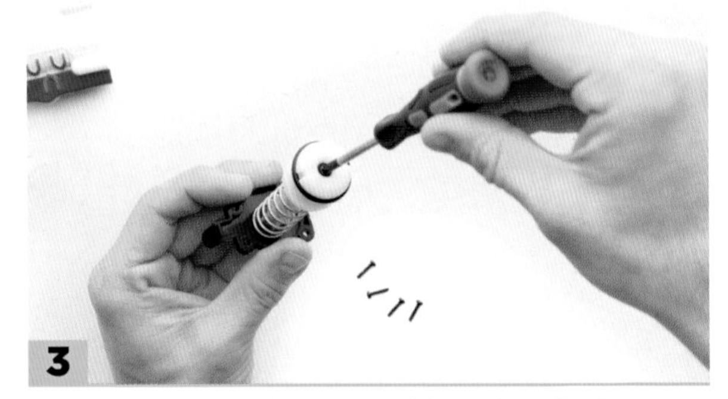

Next, unscrew the plunger head from the tube by removing the small black screw in the center.

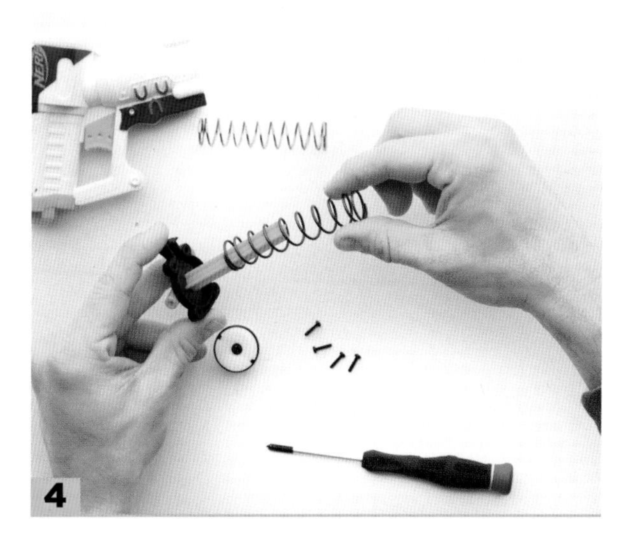

4

Remove the stock spring and set it aside. Replace it with the new spring.

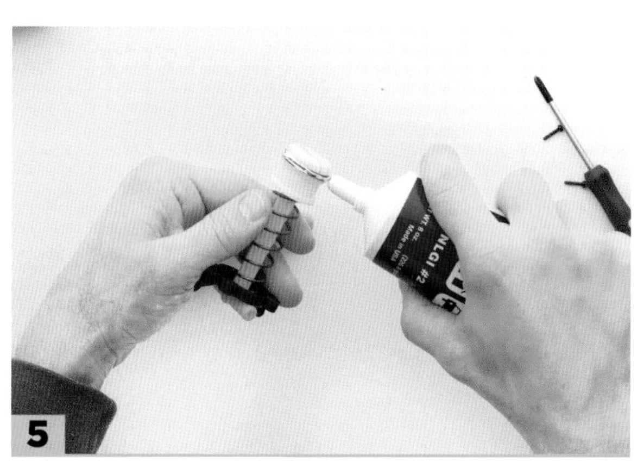

5

You can lubricate the O-ring to improve the seal between the O-ring and the plunger tube. I recommend white lithium grease or Super Lube. Other lubricants can damage the O-ring.

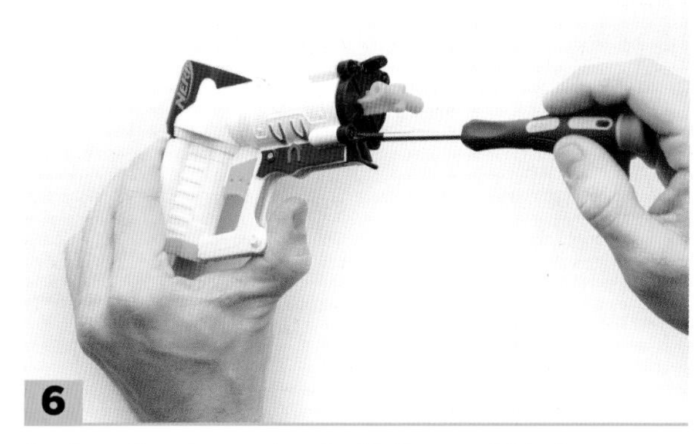

6

Replace the plunger head with the small black screw. Replace the four screws in the base.

7

Test the blaster to make sure the trigger catches and the blaster fires properly. If it does, you've successfully completed your first mod! If there are any issues, reopen the blaster and check that all the parts have been put back together correctly.

QUICK TIP

Do not store darts inside a blaster! The Triad has a "smart air restrictor" that chooses which dart to fire. Leaving the darts inside wears out the air restrictor springs and can ruin the blaster.

Time: 10–20 minutes

NERF RIVAL KRONOS

TOOLS AND MATERIALS

- #0 screwdriver
- White lithium grease or Super Lube (optional)
- Kronos upgrade spring or K26 spring
- Bolt cutters (if you intend to cut your own K26)

The Kronos along with the tools and materials for the mod.

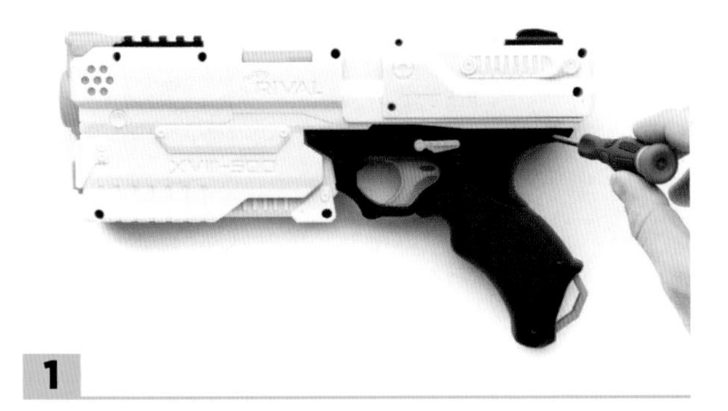

1

First, open the blaster: use the screwdriver to loosen all 14 screws on one side. It's easiest to leave them in place. Don't take them completely out of the shell.

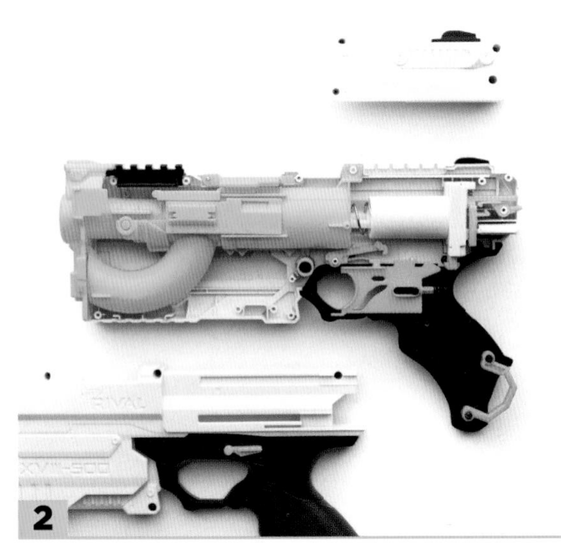

2

When all the screws are loose, pull off the top slide and shell. Keep the screws in place and set aside the slide and shell.

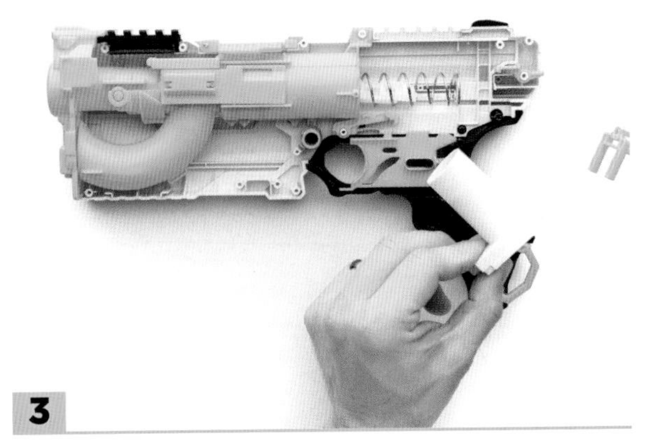

3

Remove the orange plastic catch, white plunger housing, spring, and white plunger tube.

4

You can lubricate the plunger tube O-ring for a better seal. I recommend white lithium grease or Super Lube. Other lubricants can damage the O-ring.

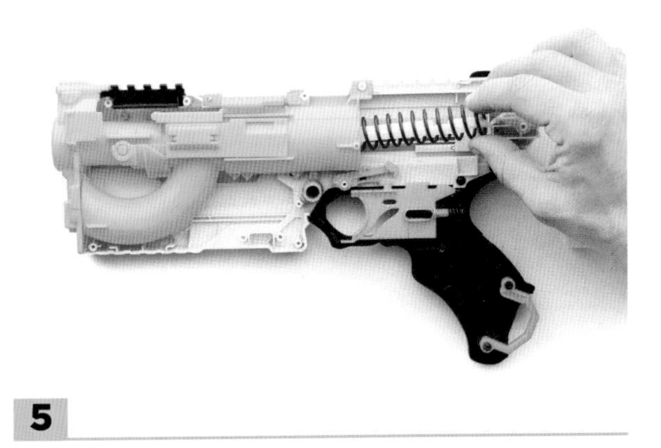

5

Remove the stock spring and replace it with the thicker K26 spring.

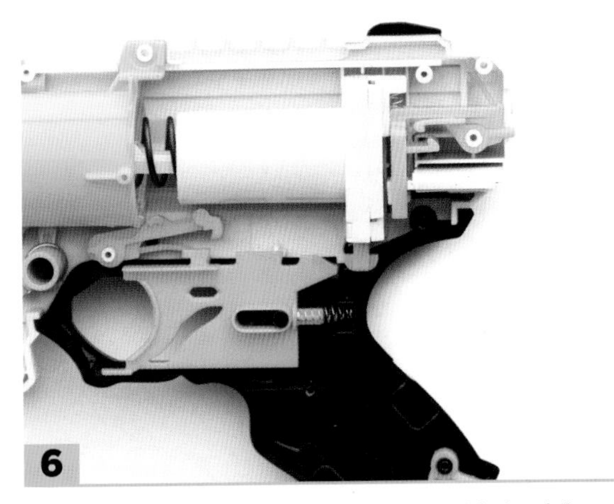

6

Place the catch and rear plunger assembly back in place. Compare your blaster to this image for proper part placement. If you would like to be able to double-fire your blaster (shotgun blast), remove the three-sided orange piece that sticks out the back of the blaster. You will lose the priming indicator by doing this part of the mod.

7

Put the shell back on the blaster and close it up, replacing all the screws.

8

Replace all the screws. Ensure the top jam door is secured in place.

9

Next, replace the top slide and the four screws. The bottom left screw is the tiny one. The others are all the same size.

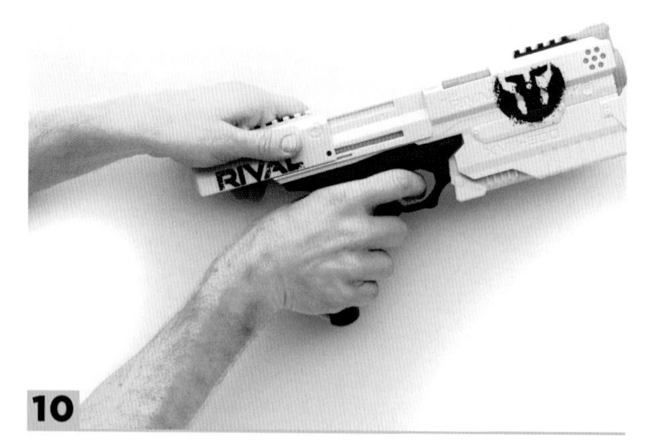

10

Test-fire the blaster. If everything catches, you have successfully modded your Kronos blaster! If not, work backward through all the steps and make sure each piece was correctly reassembled. When in doubt, return the blaster to stock form. This blaster has a very heavy prime with a K26 spring, but there are lightweight options for a more comfortable prime. Remember, more power always means a stronger prime.

Time: 20-40 minutes

NERF RECON MKII

TOOLS AND MATERIALS

- #0 screwdriver
- Tape (optional)
- Hot-glue gun (optional)
- White lithium grease or Super Lube (optional)
- Blasterparts Recon MKII upgrade spring kit (or other Recon MKII spring)

The Nerf Recon MKII mod is the most complex of our three springer upgrades.

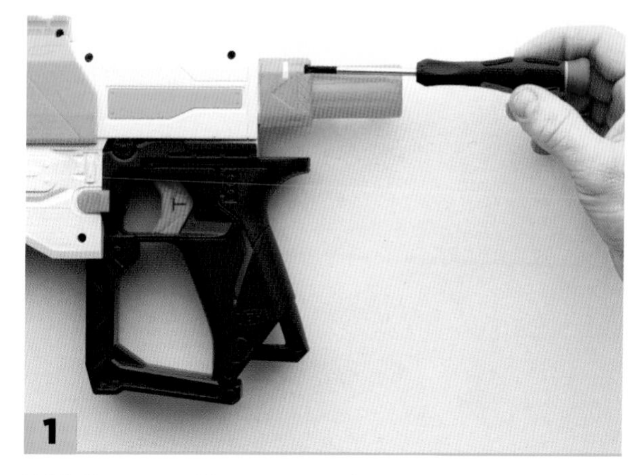

1

With the screwdriver, remove the two screws in the rear orange cap. This is where the plunger retracts when the blaster is primed. Remove the cap and set it aside, along with its screws.

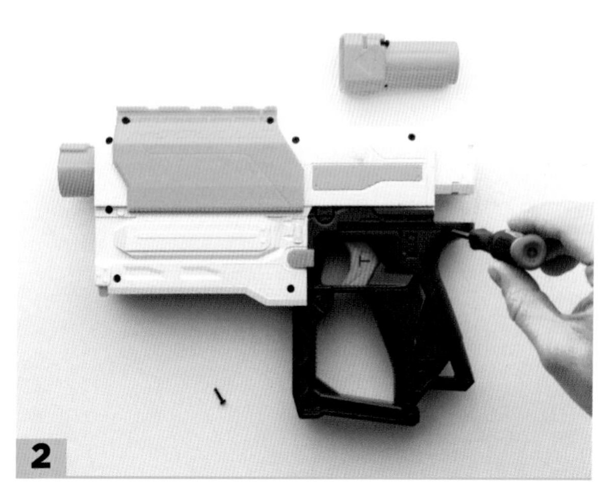

2

Next, remove all 12 of the main screws. You can tape these in place in the shell or set them in a bin. The shortest ones are for the top rail, but don't worry too much about mixing up the sizes. Lose a screw? You can always find these online, and they are inexpensive.

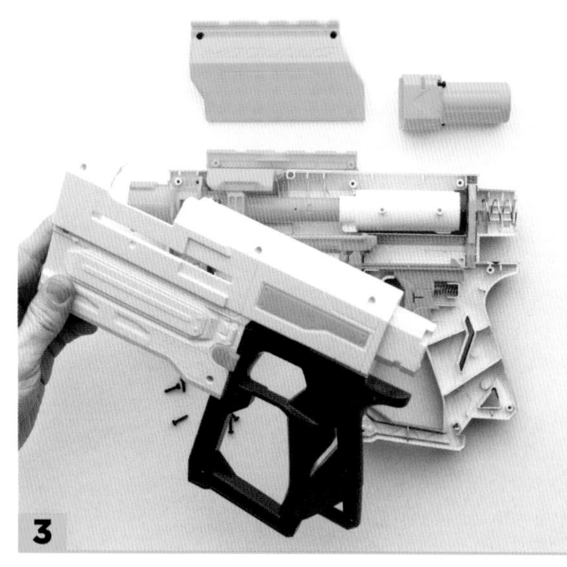

Remove the top half of the priming slide (the lighter orange part), then lift the shell off. The jam door on top may need to be opened to remove the shell. Be careful: the trigger spring loves to go flying! If you have a hot-glue gun, you can apply a tiny dab of glue on the front of the spring where it attaches to the trigger.

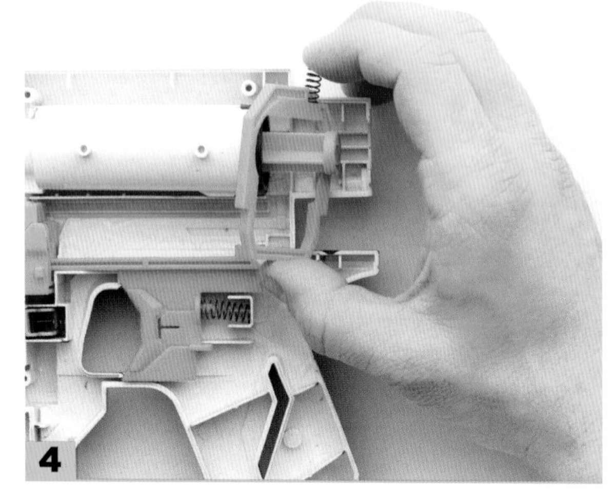

Next, pull up the trigger catch and remove the spring. Set it aside.

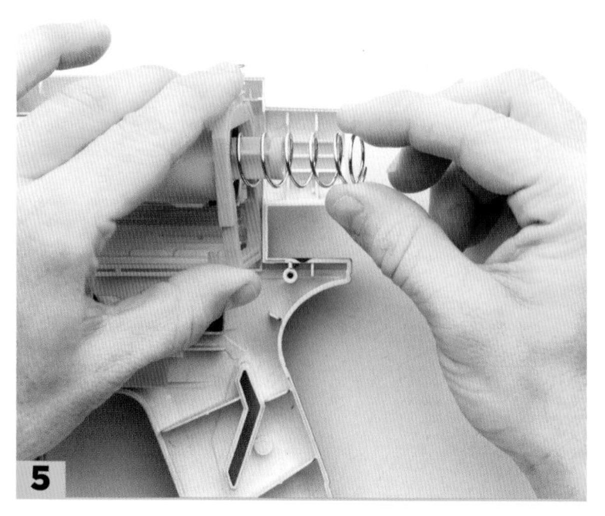

Now remove the plunger tube from the back of the blaster, along with the trigger catch and spring.

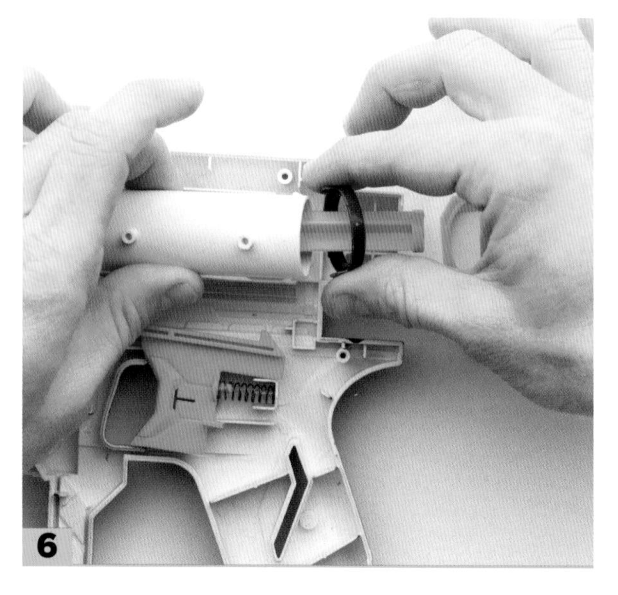

To open the plunger tube for lubrication, you'll need to pry the small gray tabs on each side of the plunger tube. Do this carefully. They are delicate and easy to break off. Use your fingernail or a small screwdriver. If you break these tabs, the cap can be glued back in place with Super Glue or hot glue.

7

If you wish, lubricate the plunger tube O-ring for a better seal. I recommend white lithium grease or Super Lube. Other lubricants can damage the O-ring.

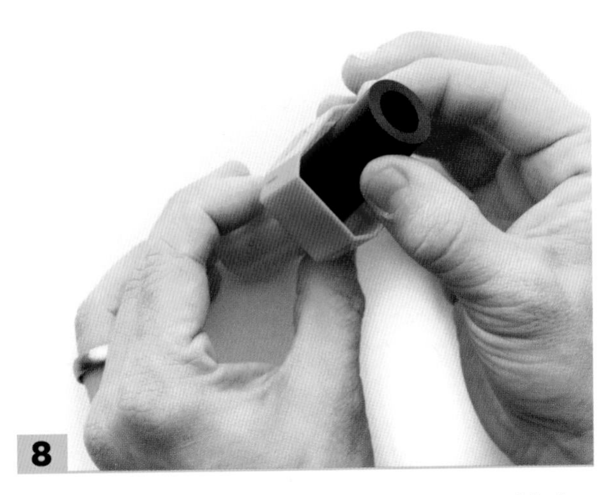

8

Insert the plastic reinforcement in the rear cap with the flat side facing forward. It simply slides in and is very easy to install.

9

Here's what the plastic reinforcement should look like inside the cap.

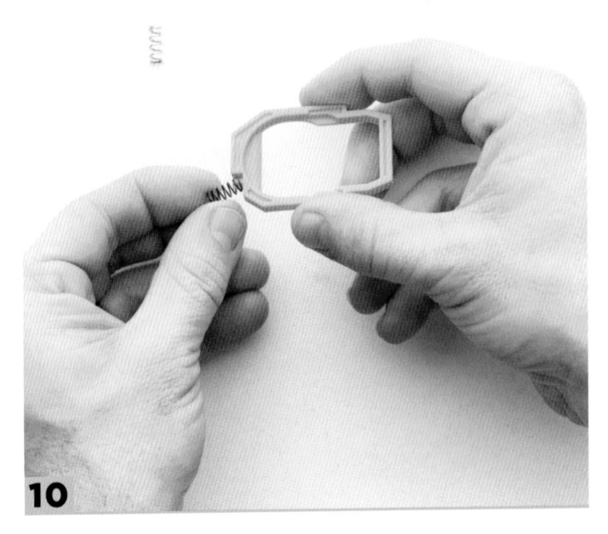

10

Now we replace the lightweight catch spring with the heavier one that came with the kit. Not all springs require replacement, but it's good to confirm when upgrading a spring.

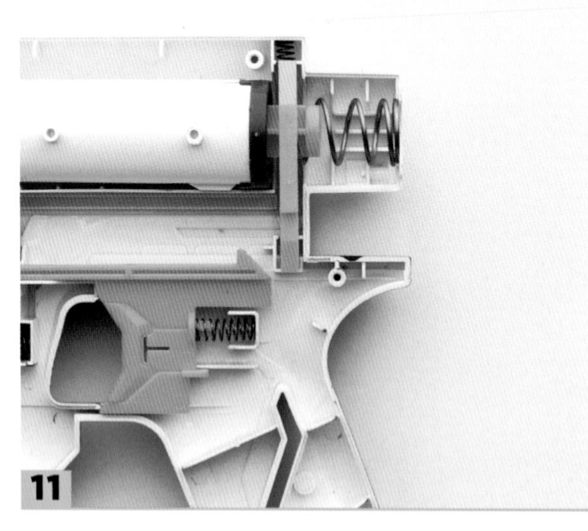

11

Next, we replace our main spring with the heavier upgraded one.

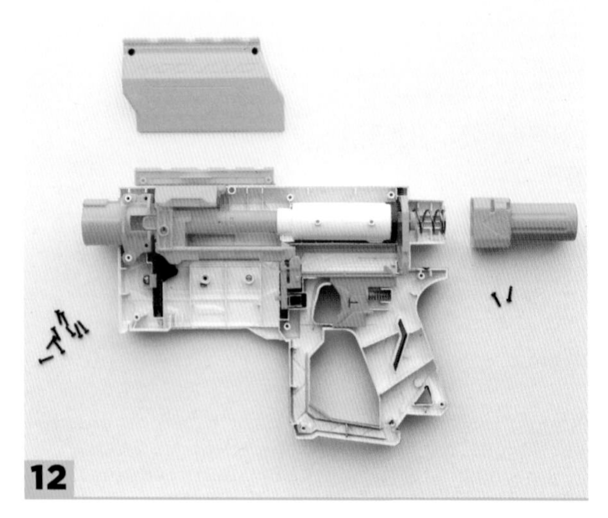

12

It's time to reassemble the blaster in reverse order.

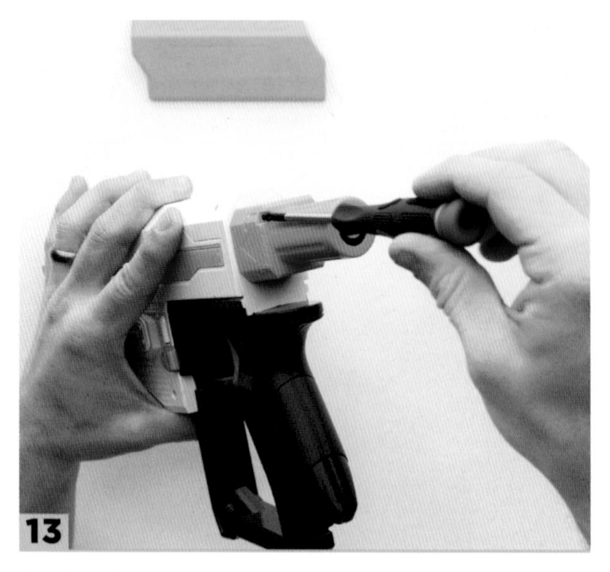

13

Replace the rear cap with the two screws that hold it in place.

14

Replace all the shell screws as well as the top slide, screwing it in place.

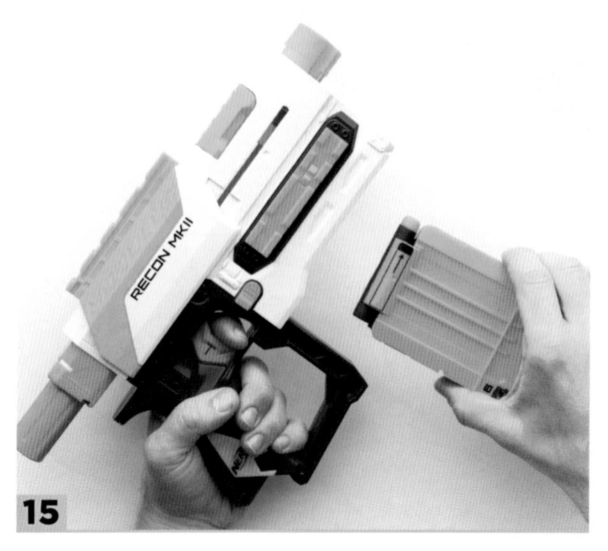

Load the blaster with a full magazine for testing. This blaster can have the magazine removed only when the slide is in the back position. This is because of how the plunger tube moves inside the blaster. Players new to this blaster often struggle with it. It's nice to let friends know it functions before a game.

Test the blaster by priming it to confirm that the trigger catches and the blaster fires properly. Congratulations! You've completed the Recon MKII mod. The stock blaster fired around 60 fps, while your modded one can fire 90+ fps.

Voltage, Resistance, and Current

When you learn about electricity you will hear a lot of new words. Here are three that describe the most basic concepts of electricity.

Voltage: This is a measurement of electromagnetic force. Think of voltage like water pressure in a tube. If you increase the pressure, more water will flow through the tube. Voltage is measured in volts (V).

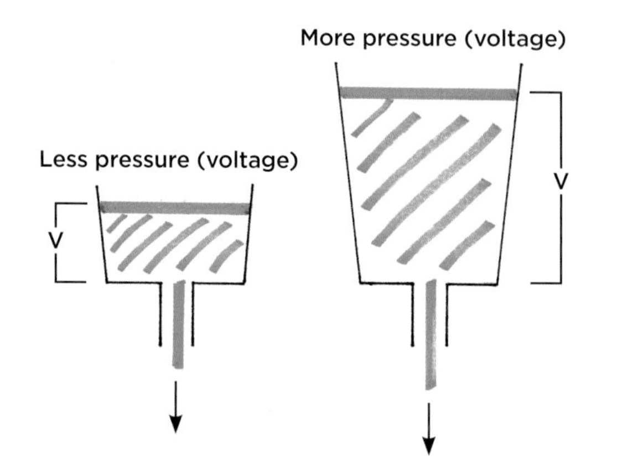

Voltage can be compared to the force with which the water tank pushes water through the pipe. The taller tank on the right has more pressure and can push more water through the pipe.

Resistance: This is anything that prevents electricity from flowing freely through your circuit. Wires, switches, and other parts all add resistance. That's why it's important to use the right parts in your build. The more electricity your motors use, the heavier wire and switches you will need.

Think of resistance as the size of the water pipe. The larger, longer, or thicker the pipe, the more water can flow through it. Larger wiring does the same thing for electricity. Have you ever drunk water from a long silly straw? How much easier is it to drink through a normal straw? The longer silly straw has more resistance.

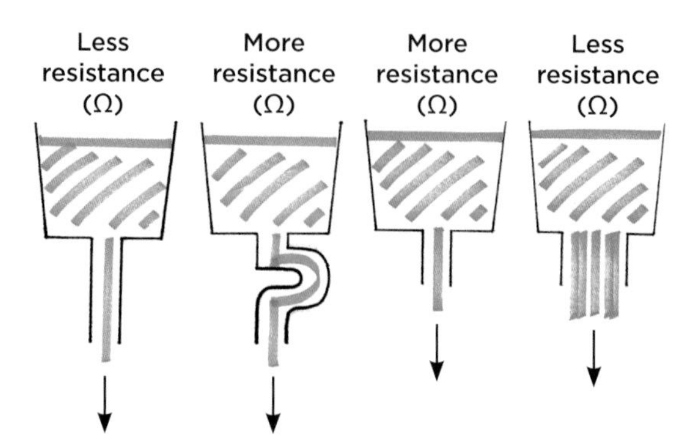

Which of these water pipes will provide the most resistance? Just like the water pipes, longer or thinner wires offer more resistance. This means we never want to use extra wire without a good reason.

Current: This is the rate at which electricity flows through a circuit. Current is like the flow rate of water in pipe. It is directly related to how hard the water is pushing (voltage) and how large the pipe is (resistance). Current is measured in amps (A).

So, we want to match our wiring, switches, and battery to our motors (though larger wiring will never hurt). We also have to match our battery voltage to our motors. Using a battery that gives too much voltage can burn out motors.

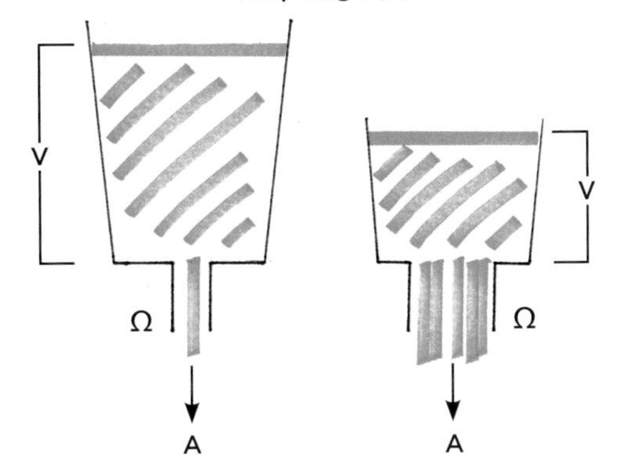

The combination of voltage and resistance gives us the flow rate (current or amperage).

The wrong switches and wiring not only rob motors of power, but they can also cause the circuit to fail. A battery that's too weak can also damage a circuit. In the case of LiPo batteries (see Chapter 2), it can be dangerous too.

Batteries

A battery stores energy. Inside it has two elements. When the elements combine, they cause a chemical reaction that produces electricity.

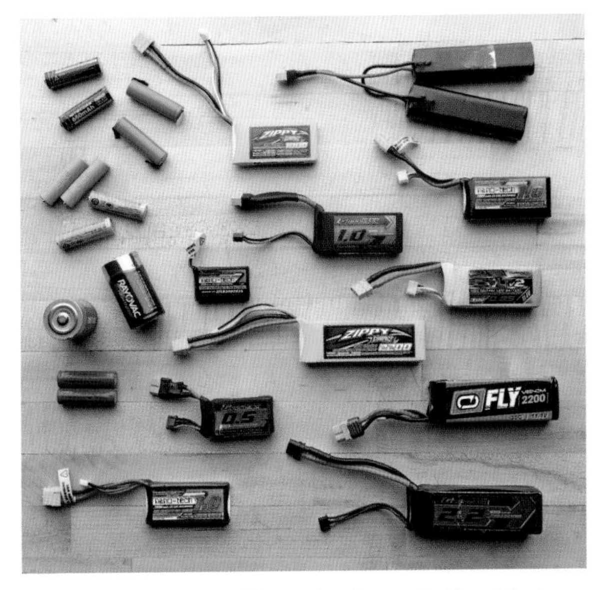

Here's a wide range of batteries, from alkaline AAs to high-powered LiPo batteries.

WIRE GAUGE

AWG (American Wire Gauge) measures the "gauge," or thickness, of wire in the United States. Electricity flows more easily through thicker wire. It gets a little tricky because a larger AWG number describes smaller wire. So, 16AWG wire is thicker than 22AWG.

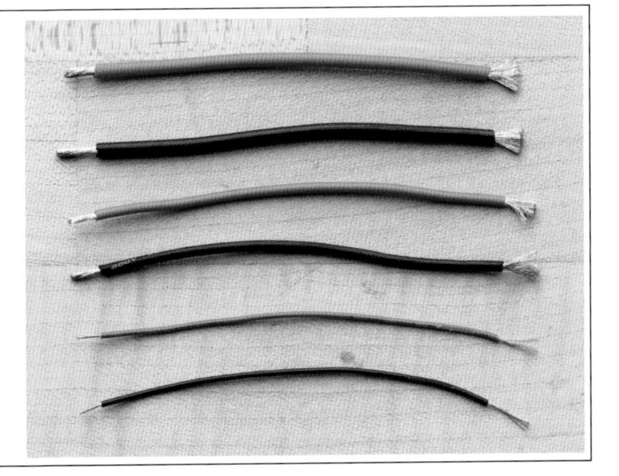

You will hear about four main kinds of batteries in Nerf blasters: alkaline, NiMH rechargeable, IMR lithium cells, and LiPo.

Alkaline (zinc manganese): Stock blasters from the store use disposable alkaline AA, C, or D batteries. Alkaline batteries cannot be recharged. These batteries are expensive over time. They are also bad for the environment because they are usually thrown out after they "die." Alkaline batteries also offer poor performance compared to other kinds. Upgrading an electronic blaster with rechargeable batteries can save money over time because most can be recharged hundreds of times.

NiMH (nickel-metal hydride): NiMH batteries are used in a few Nerf blasters, including the Rival rechargeable pack, TerraScout, TerraDrone, and Rival Prometheus. NiMH batteries are very safe and simple to use. But they don't provide as much power as lithium batteries. NiMH batteries are also larger than lithium batteries.

IMR (lithium manganese): IMR, or lithium, batteries are available in many sizes, including the shape of AA cells (called 14500). These batteries are often used to replace stock alkaline batteries in entry-level mods. They are often of very poor quality, but there are some high-quality options, such as the yellow Nitecore brand 6.5A batteries on the previous page.

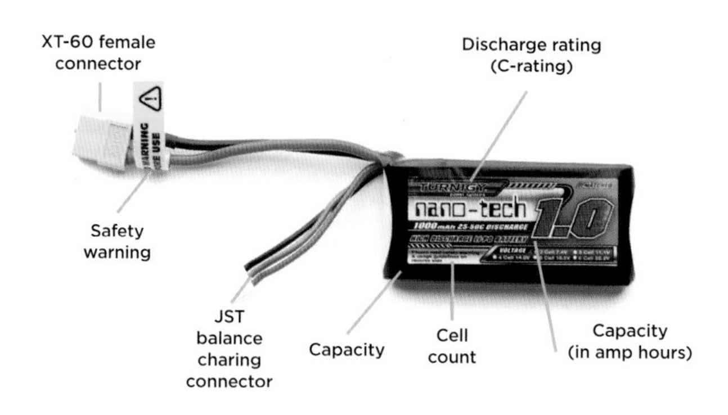

LiPo Battery Label Guide

If you install these in a blaster, you must use the blaster's stock battery terminals, which can only handle a small amount of electrical current.

IMR batteries are much safer than LiPo batteries because manganese produces less heat. Some even have protective circuits that control the battery. I suggest using IMR batteries only with the stock motors because almost all upgrade motors need too much current to use IMR batteries.

LiPo (lithium polymer/lithium ion): These are the ultimate batteries for Nerf blasters. They also require that you use them with caution. Read about LiPo safety in Chapter 2 before using them, and always read and follow any warning labels.

LiPo batteries can provide very high current and capacity in a small size. Capacity is measured in Amp-hours (Ah) or milliamp-hours (mAh). 1Ah (or 1,000mAh) means that a battery can provide 1A of current for 1 hour. They come in hundreds of sizes, capacities, and current ratings. Larger packs the size of your fist can even have enough power to start a car engine!

Battery cells are individual batteries that make up a battery "pack." Cells are combined to get the desired voltage. Cell count in a LiPo battery is referred to with a small letter s. A 2s battery has two cells, a 3s battery has three cells, and so on. The number of cells determines the pack's voltage. For example, a 2s battery is 8.4 volts (4.2V x 2 = 8.4). Remember to always match your battery to your load (motors). If you try to simply swap in motors with higher voltage, they will burn out quickly.

The C-rating is a pack's maximum available current based on the resistance of the pack. Don't worry if this sounds confusing. The math is fairly simple. The C-rating is multiplied by the Ah to get the pack's available current.

Examples:

1,000mAh, 3s, 35C battery
1,000mAh* = 1.0Ah
1 x 35 = 35A (amps)

2,200mAh, 3s, 40C battery
2,200mAh = 2.2Ah
2.2 x 40 = 88A

1,000mAh (milli amp-hours) =
1Ah (amp-hour)

This amp rating is the maximum current under ideal conditions when the battery is new. As the battery gets older, so does its resistance. Resistance lowers the battery's current and voltage. This is why it's best to "storage charge" your batteries between uses (see Chapter 2).

Motors

Brushed motors are the most common kind used in Nerf. Metal brushes attached to a shaft inside the small can-shaped motor spin very fast. This creates an electromagnetic force that pushes against magnets inside the motor. This happens hundreds of times per second. It also causes the sparks you might see inside a motor.

Here are some of the motors available for Nerf mods. Those with names or labels were all custom ordered and manufactured for Nerf.

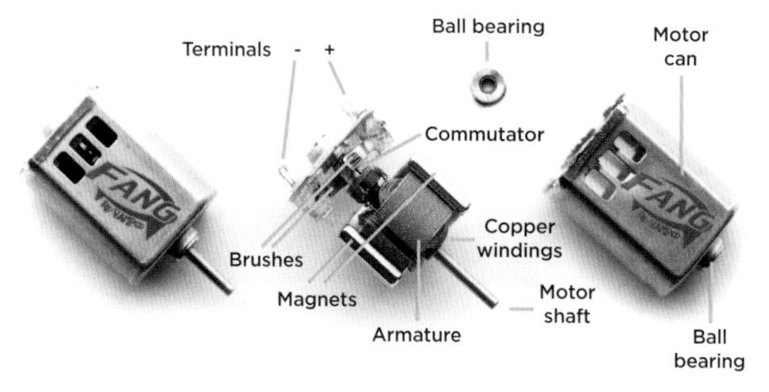

Here are the parts of a common stock Nerf motor.

Motors are available that produce a variety of RPMs (revolutions per minute), torque, and voltage. Torque is a measure of the force of the shaft's rotation (spinning). It is commonly listed in kg/cm or g/cm (kilogram-centimeters or gram-centimeters). This is how much force the motor can put out 1 centimeter away from the shaft. Motors with more torque will spin up faster at the start. They will also recover faster between darts fired.

All Nerf motors will spin equally as fast in forward or reverse (clockwise or counterclockwise). The direction depends on how you connect the battery's positive and negative wires to the motor. This is important, because two flywheels need motors that will spin them in opposite directions. It's a good idea to test motor spin direction before soldering the connection.

Motors are designed for a specific voltage. It's important to use them at their intended voltage for the most reliable results. Motors in Nerf blasters will have a 2s or 3s rating. Match them with their corresponding LiPo battery.

HOW TO GET MORE PERFORMANCE

Electric flywheel blaster performance can vary widely. It is affected by many factors: the motor, flywheel cage, flywheels, battery, wiring, switches, and darts. Changing any one of these parts can change the fps and range. Most websites that sell parts will offer basic advice on a good starter setup. It's easiest to start with a good combo that's been tested by other Nerfers.

Stock motors, while designed for 6 to 9 volts or less, can often take a higher voltage and a battery pack with a higher current. This is true for the entire Rival line. These blasters run well on 2s and 3s motors, although 3s motors will decrease the lifespan of the motors and flywheels.

The distance between the motor shafts is usually how we measure crush. It can also be measured as the distance between the two flywheels, because flywheels can vary in size. The left cage is a 41.5mm sizing. The right is a 43.5mm. That 2mm distance makes a huge difference in performance and reliability.

Flywheel Cage Crush

As you get into Nerf modding, you might hear people talking about "crush." It's really quite simple: crush is the space between the flywheels. The smaller the space, the more the dart is pinched (crushed) between the flywheels. This generally results in higher fps.

Crush can be adjusted in two ways. First, you can install a flywheel cage with closer spacing of the two motors. Open Flywheel Project 3D-printed and CNC cages come in spacings from 41.5 millimeters to 43.5 millimeters. You can also change the crush by changing flywheels. Worker, for example, makes black "high-crush" and white "standard-crush" flywheels. The high-crush wheels give 1 millimeter more crush. You can get about the same results by using a smaller crush cage and smaller wheels.

Beware that some darts with harder tips can't make it through smaller crushes. Smaller crushes also require more powerful motors and batteries and wear down motors faster.

Soldering

If you haven't soldered before, don't skip this lesson!

Soldering can be the most intimidating part of a Nerf mod, but it's worth learning. A proper solder joint will last forever.

So what *is* soldering? Soldering is joining together two parts of an electrical circuit. This can include connecting a wire to other wires, switches, motors, battery connectors, LEDs, and dozens of other parts.

First, the two parts to be joined and the solder (pronounced SOD-der) are heated with a special tool called an iron. Then the solder, a soft metal, is added. The iron melts the solder, which flows around the point where the two parts are joined. The joint is held in place. Once the solder cools, the joint should be permanent.

These are flywheels for darts (left) and Rival (right). The flywheels cause a lot of vibration when they spin at high speeds.

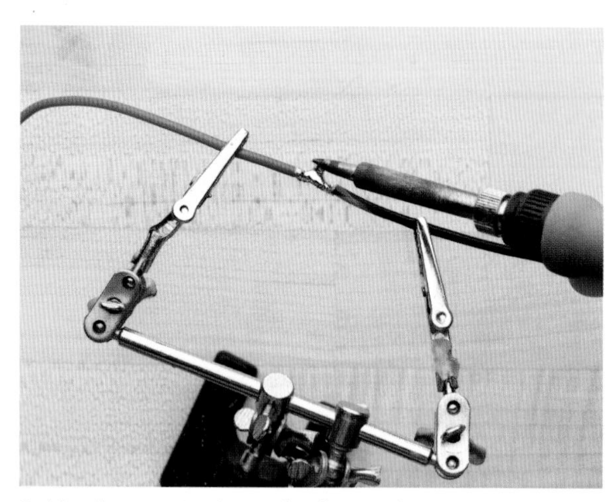

Solder flows onto the ends of two wires.

It's important to heat the wire and the solder at the same time. You should also keep everything as still as possible while it cools. If a solder joint is moved while cooling, you can end up with a weak joint. It might pull apart easily. It might also create more resistance to electrical flow. Our solder joints need to be strong because many Nerf blasters vibrate a lot.

Soldering is all about practice and patience. That's how I got much more confident. I used to avoid soldering when I started. With a little practice, you can easily get over this fear just like I did.

For those just starting in Nerf modding, I highly recommend practicing on a few pieces of spare wire before soldering inside a blaster. In the following activity, we will learn how to solder together two pieces of wire.

SOLDERING PRACTICE LESSON

TOOLS AND MATERIALS

- Safety glasses
- Sponge (a clean, unused kitchen sponge will do)
- Spray bottle
- Wire strippers or hobby knife
- Spare wire
- Helping hands (optional)
- Heat shrink (optional)
- Soldering iron
- Solder
- Heat gun or hair dryer (optional)

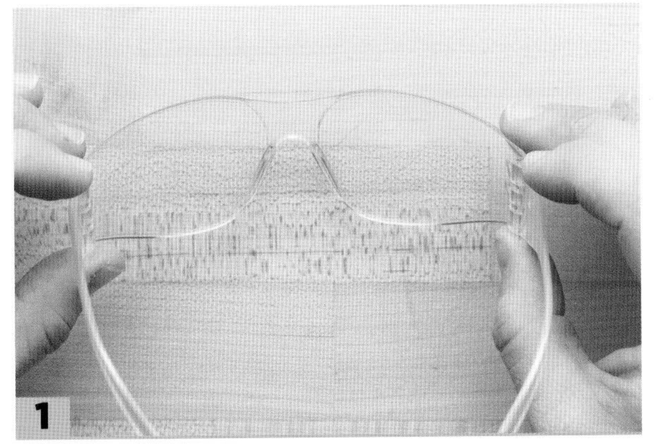

1 Always wear safety glasses when soldering!

2 Wet your sponge with a spray bottle. It doesn't have to be dripping wet, just very damp.

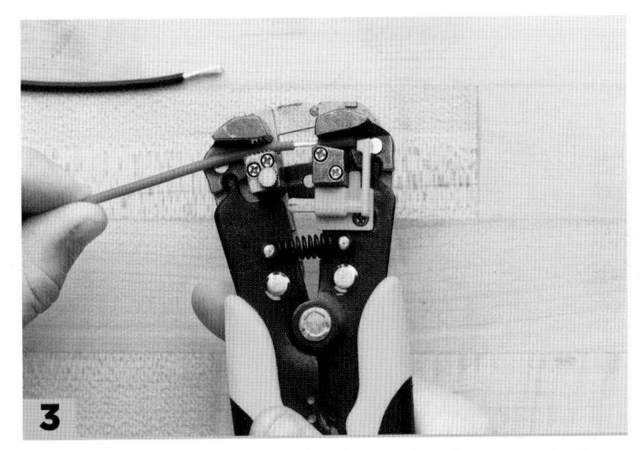

3 With wire strippers, strip the insulation from the last ½ inch (13 millimeters) of each wire.

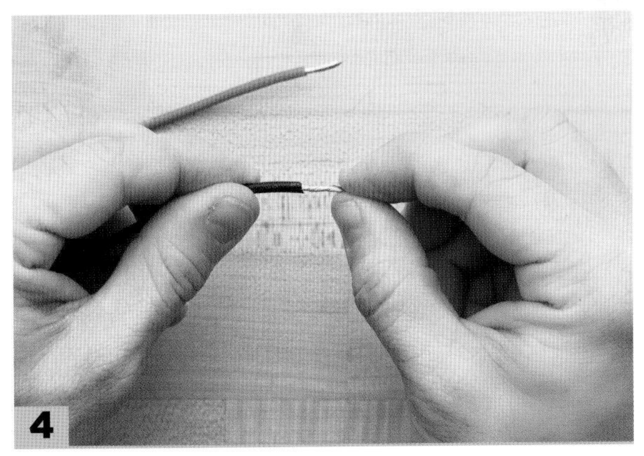

4 Twist the strands of wire together and use helping hands to hold them in place. You can also set them on the table, as long as you don't mind getting a bit of hot solder on it.

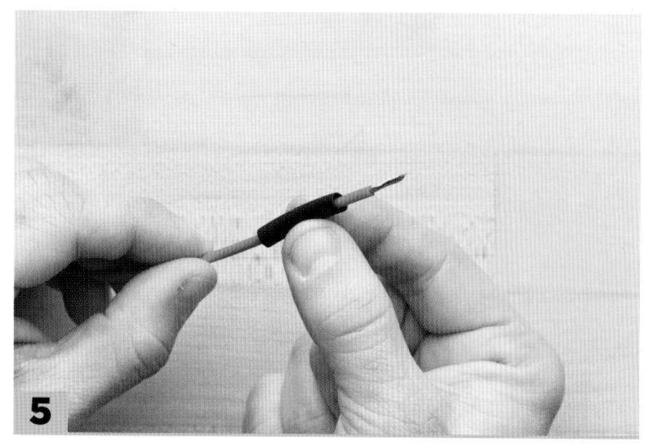

Slide a piece of heat shrink down the wire, out of the way of where you will be soldering. It's helpful to get in the habit of adding heat shrink *before* soldering. More often than not, you can't add it later.

Tin your soldering iron by applying a bit of solder. Roll the tip of the iron around. This allows the bead of solder to clean the tip. Then wipe it off on your wet sponge.

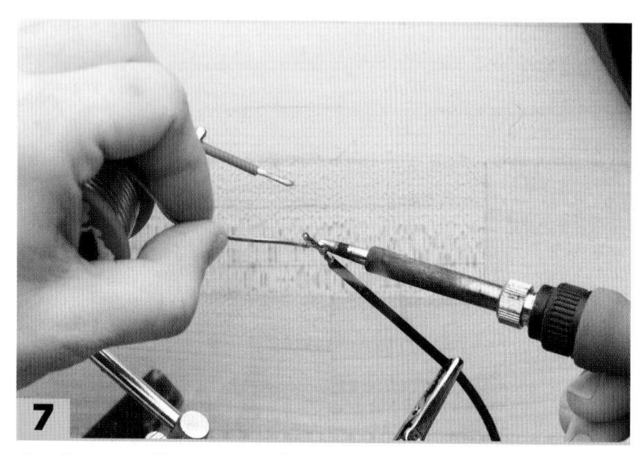

Apply a small amount of solder to one wire to be connected . . .

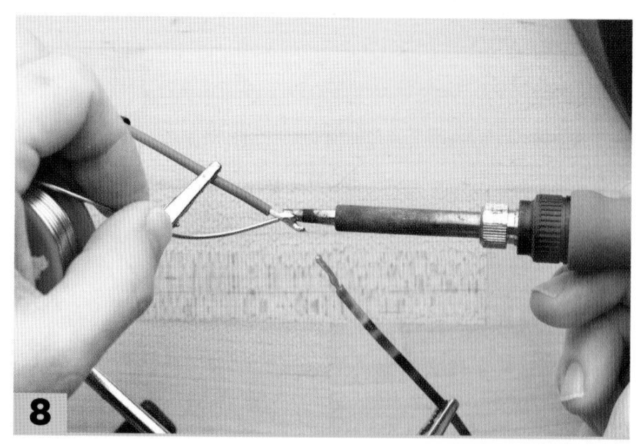

. . . then to the other.

After each solder joint, wipe the iron tip on the sponge to remove excess solder.

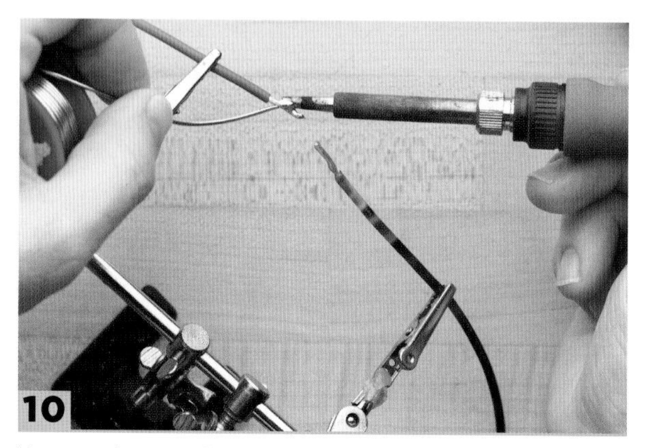

Now apply a small amount of solder to the iron tip. Hold it steadily on the two wires until the solder flows over the entire joint.

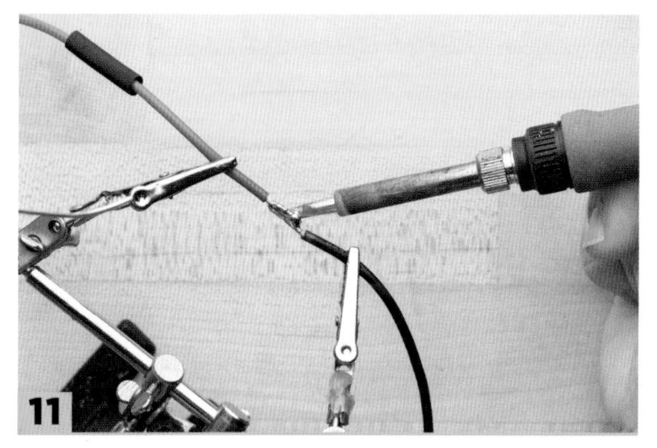

When the solder is liquid-like, we say it has "fluxed." If all the exposed wire has solder on it, you're ready to stop.

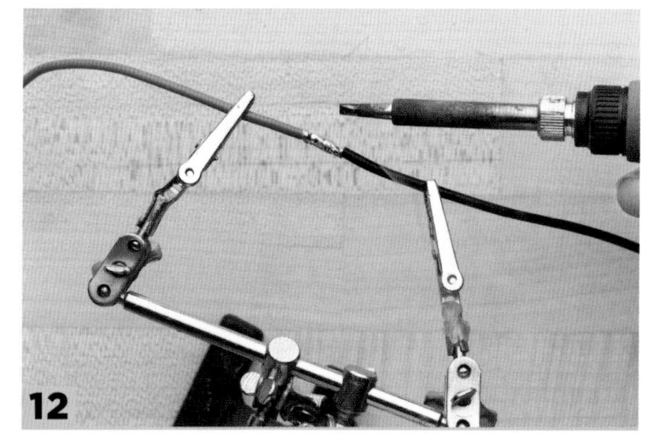

Remove the iron tip from the wire. Be careful to move the soldered joint as little as possible. Wait 5 to 10 seconds for the solder to cool.

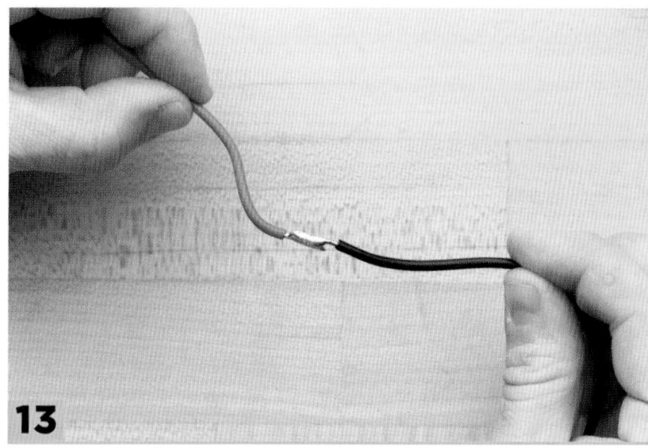

Test the strength of the joint by tugging at the wire. It shouldn't come loose when pulled.

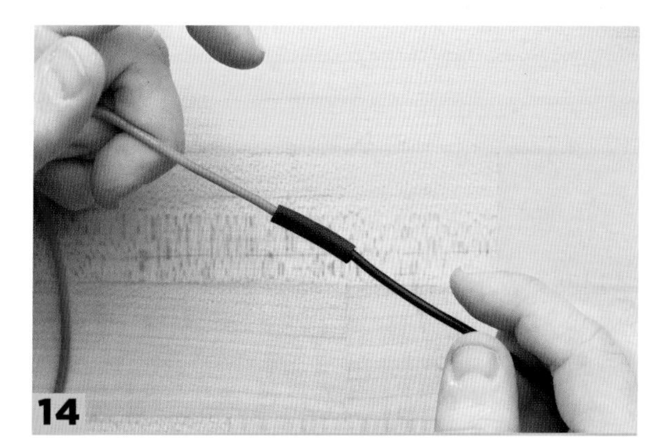

Slide the heat shrink over the joint so it covers the exposed wire completely.

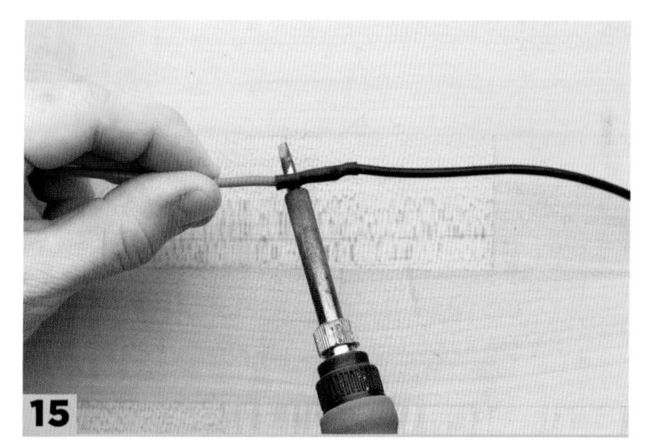

Finally, use a heat gun or hair dryer to shrink the heat shrink. You can also use the side of the soldering iron as shown.

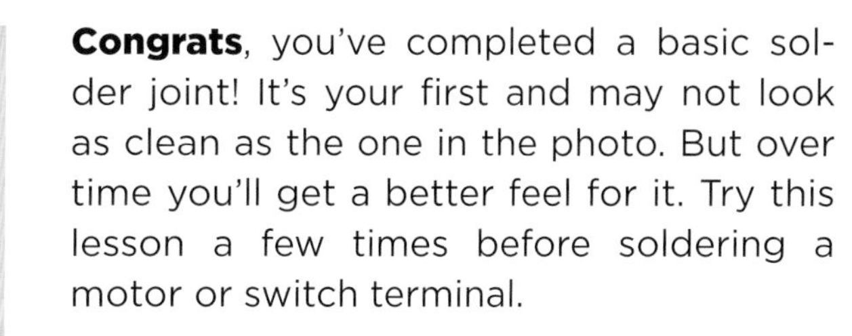

Congrats, you've completed a basic solder joint! It's your first and may not look as clean as the one in the photo. But over time you'll get a better feel for it. Try this lesson a few times before soldering a motor or switch terminal.

The Process

The process for modifying an electric flywheel blaster is straightforward. First, we open the blaster and remove all the wiring and electrical parts we don't need. Then we upgrade the wiring and replace the switch. Next, we swap in better motors and make room for a higher-powered battery. Finally, we test the blaster to confirm everything functions as expected and then reassemble.

There are several optional internal upgrades, including flywheels, flywheel cages, and motors. Finally, there are add-ons and optional paint jobs that we will cover in a later chapter.

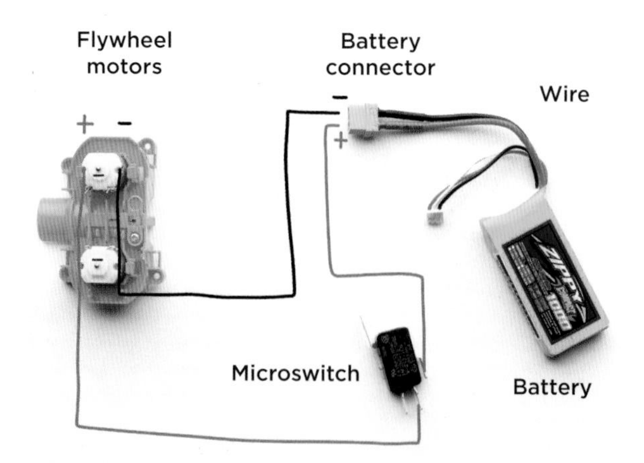

NERF BLASTERS WITH THE STRYFE WIRING LAYOUT

- Stryfe
- Elite Demolisher
- Modulus
- Rival Zeus
- Rival Hera
- Elite Rayven

Most flywheel blasters use the same wiring layout as the Stryfe. That means you can use this mod on the blasters listed here.

In this step-by-step mod, we are going to do it right. That means we'll use a high-powered LiPo battery that requires the replacement of all wiring, motors, and switches. We'll also need to expand the battery compartment.

MATERIALS

- Nerf Stryfe blaster
- 2s 1000mAh LiPo battery with a C rating of 35C or higher
- LiPo voltage alarm (required for safe operation)
- 15A or 21A long lever switch
- Epoxy putty (EP-200), hot glue, or 3D-printed switch plate
- Two Meishel 2.0 motors (or other motors of your choosing)
- 2 feet (0.6 meter) 18AWG red/black fine-strand silicone hobby wire (higher-powered motors will require 16AWG wire)
- AA, C, or D battery
- Electrical tape
- Solder
- 3⁄16-inch (5-millimeter) heat shrink
- XT-60 connector
- LiPo charger
- LiPo charging bag (for safe charging)
- LiPo voltage alarm (to monitor battery)

STEP 1: REMOVING ELECTRONICS AND TRIGGER LOCK

Now that we have all our tools and materials together on a clean workspace, let's get started.

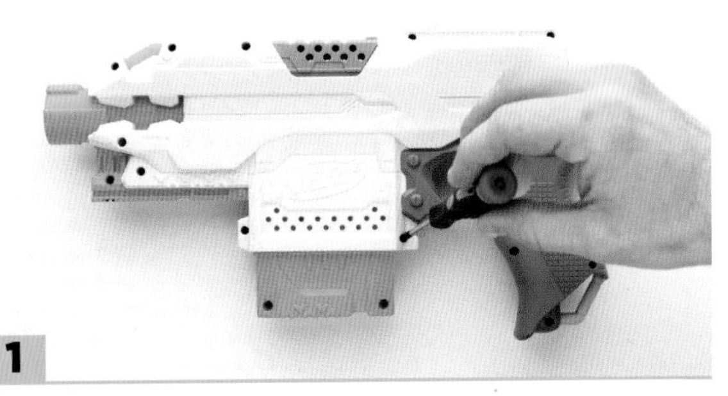

1

Put on the safety glasses. With the screwdriver, open the blaster, unfastening 13 screws. Skip the front bottom screw in the orange plastic. Leave the screws in place in the shell or place them in a small dish. Don't worry too much about mixing up the different lengths. The shortest ones are for the top rail. The rest of the blaster has identical screws.

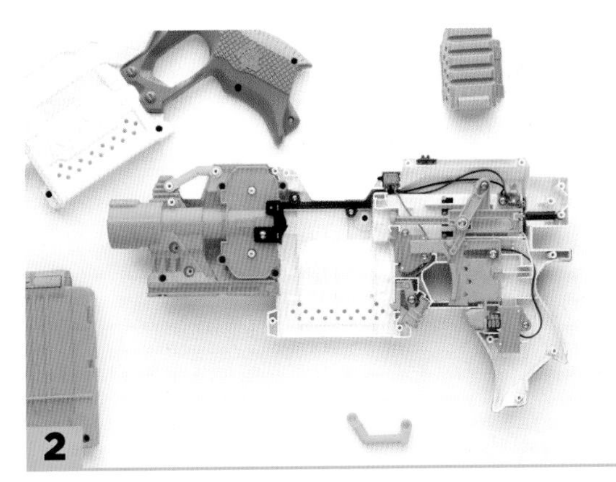

2

Remove the top half of the shell and set it aside. We will make one small mod to the inside of this shell later. Remove the magazine, top jam door, and sling mount on the handle and set aside.

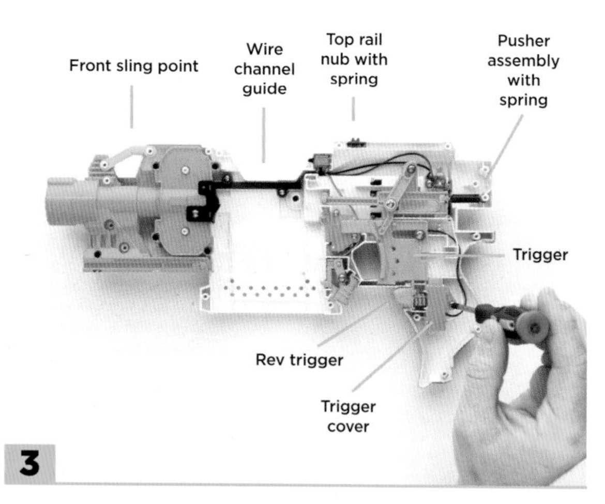

Remove the stock front sling point, trigger cover, rev trigger, wire channel guide, top rail nub with spring, trigger, and pusher assembly with spring.

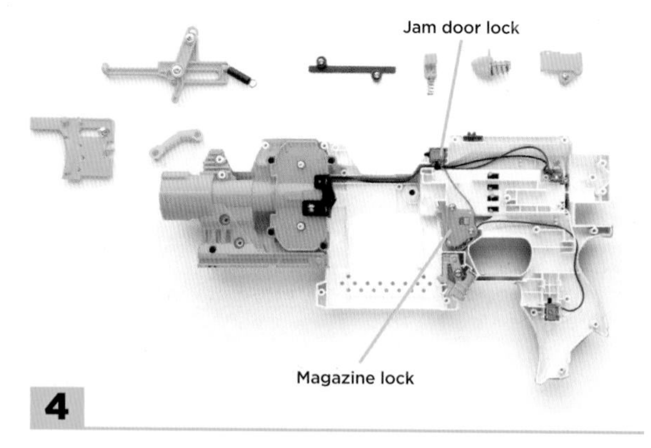

The magazine lock prevents firing when there is no magazine in the blaster. The jam door lock prevents the blaster from powering up when the door is open.

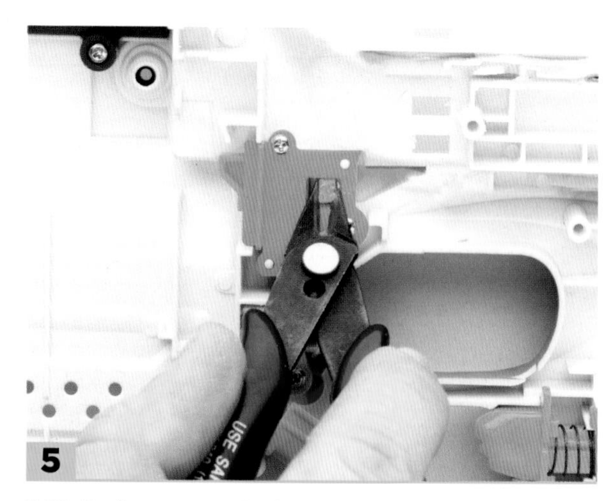

With flush cutters, snip the catch on the trigger lock. We don't need this lock, but I like to keep the spring-loaded tab. It helps stabilize the magazine when the magazine is inserted in the blaster. If you prefer, you can remove the entire assembly.

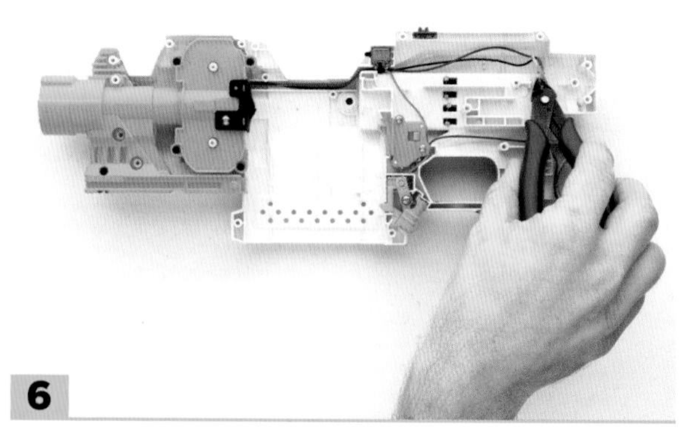

Unscrew the small circuit board. This is a thermistor that prevents too much current and potential damage to the wiring and motors. Snip or cut all existing wiring out of the blaster. There is no need to be careful here—we are going to get rid of everything electronic in the blaster. We'll remove the small switches that are safety locks. Locks are a way for Nerf to make the blaster more "kid-proof." I remove the locks completely because they interfere with operation and maintenance.

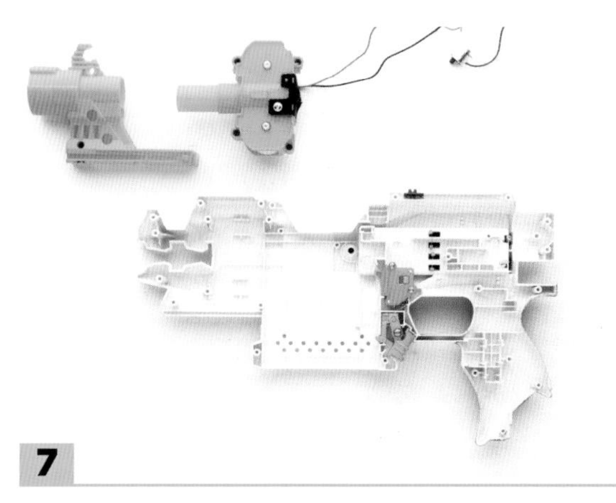

7

Remove the four flywheel screws and pull the motor, blaster muzzle, and wiring out of the blaster.

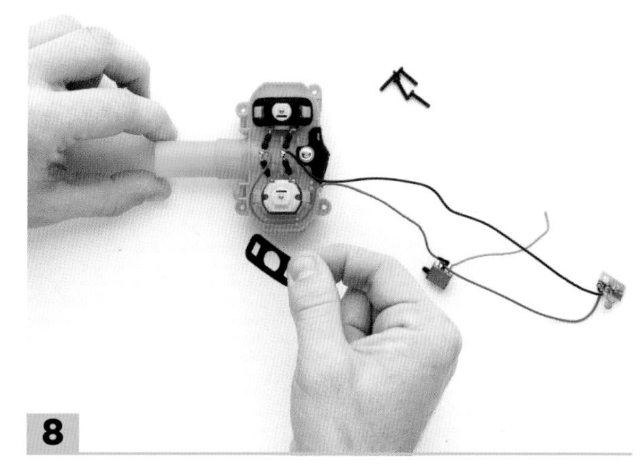

8

Remove the rubber tabs holding the motors in place.

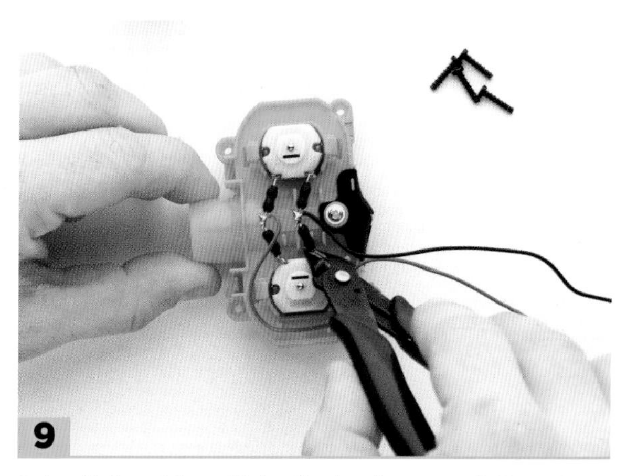

9

Snip all the wiring off the flywheel cage and discard.

10

Remove the screw holding the rubber dart skirt. Discard all three pieces. We won't need them for our upgrade.

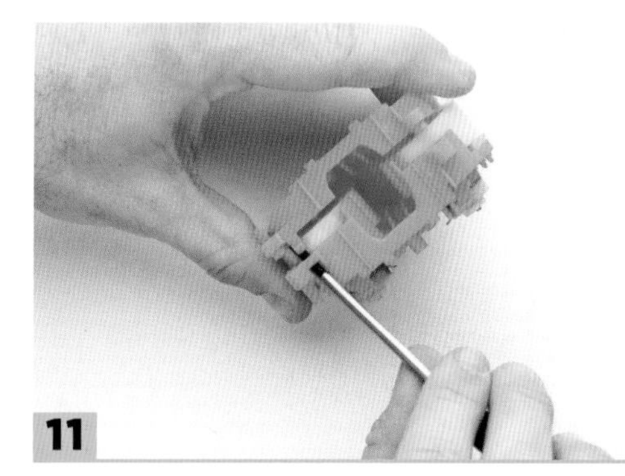

11

Use a small flathead screwdriver to carefully open the tabs of the flywheel cage.

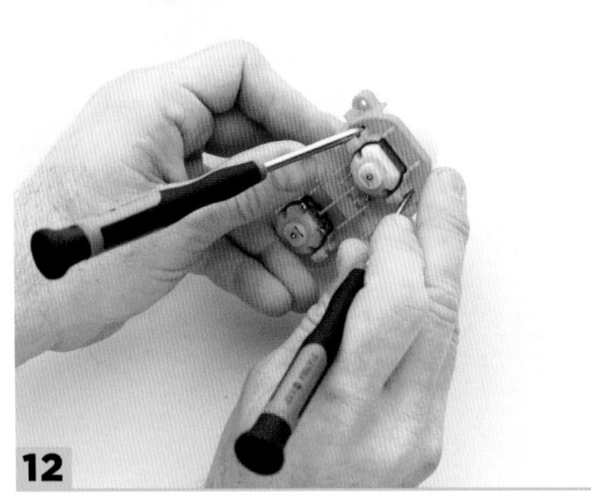

12

Next, use two small screwdrivers to push the flywheels off the motors. If you intend to reuse the flywheels, do this very gently and take your time.

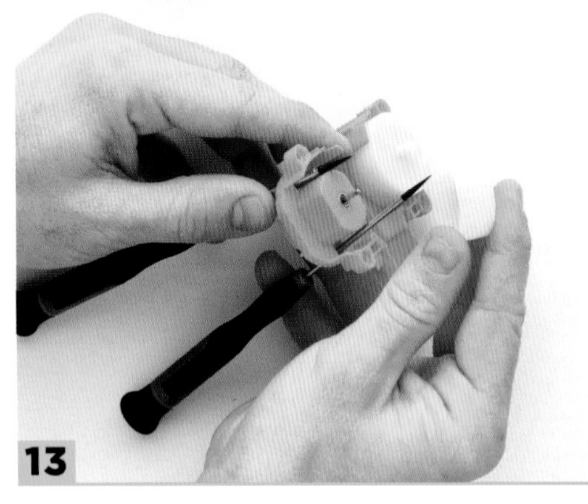

13

The flywheels have come off.

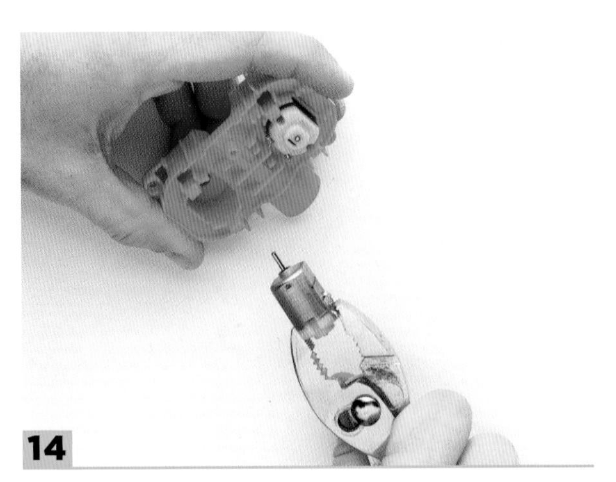

14

Using a large pliers, pull both motors out of their mounts and set aside. These motors are very low quality and low power. They will have to go!

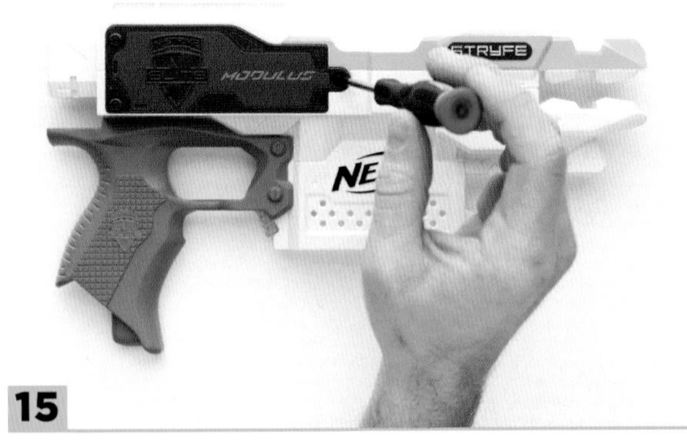

15

Remove the battery door by removing this single screw. You can purchase Stryfe thumb screws, which make this easier to access without tools after your mod is complete.

STEP 2: CUTTING AND DEMOLITION

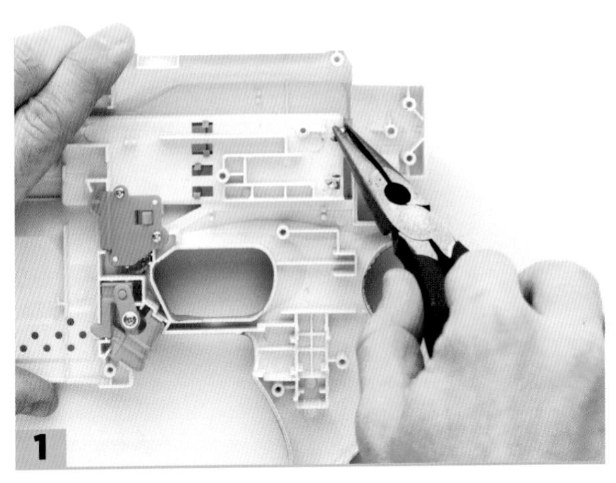

Bend these tabs for the original battery mounts. They need to be pulled out from the other side. Snip them if needed.

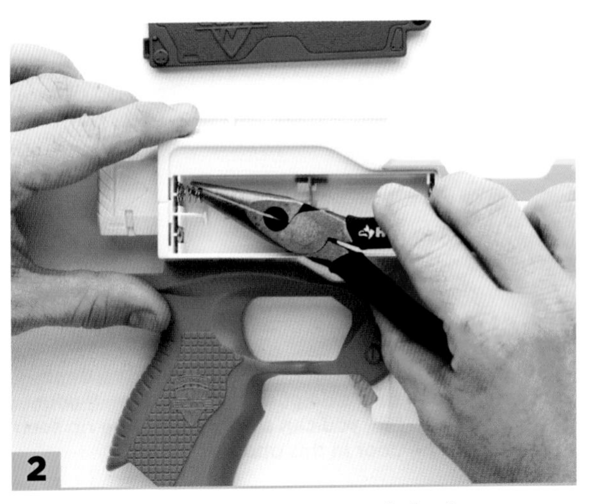

Pull and remove all four wire tabs and plastic battery spacers with a pliers. Twist and work them out completely, and then discard. Again, no need to be careful.

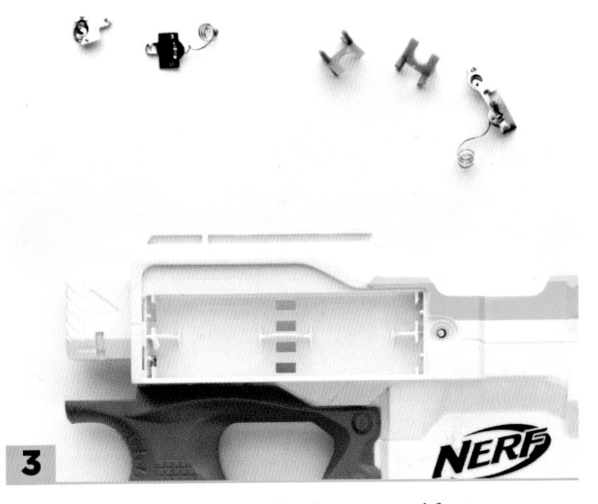

We want all unnecessary plastic removed from this compartment.

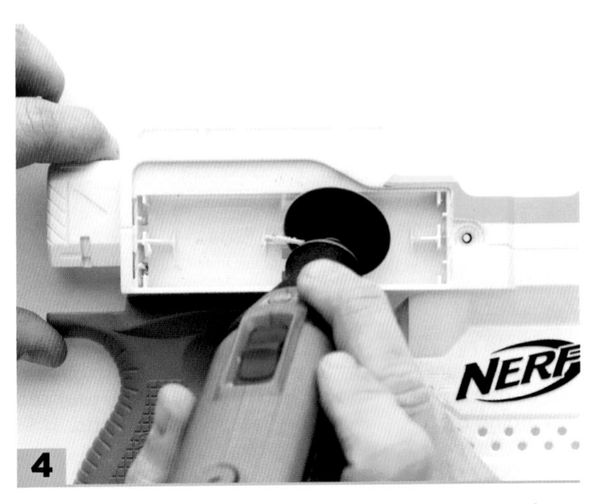

Use a rotary tool, flush cutters, or snips to remove the unwanted plastic. Sometimes it's a good idea to use a Sharpie to mark the plastic you need to remove. Test-fit your LiPo battery and voltage alarm or voltage meter as you go. You may need to cut additional space toward the front of the compartment.

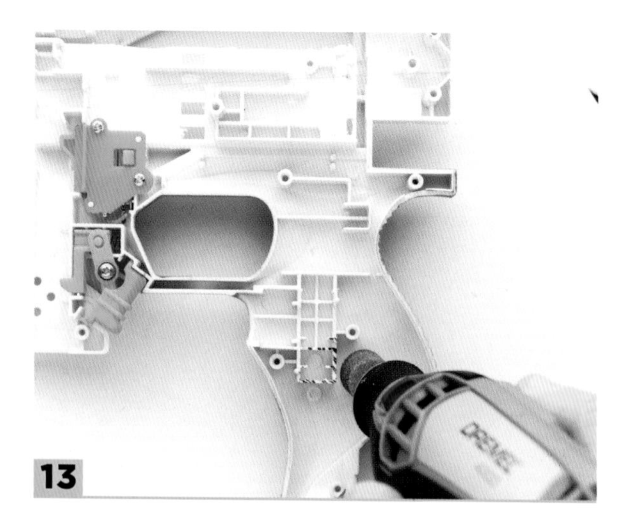
13

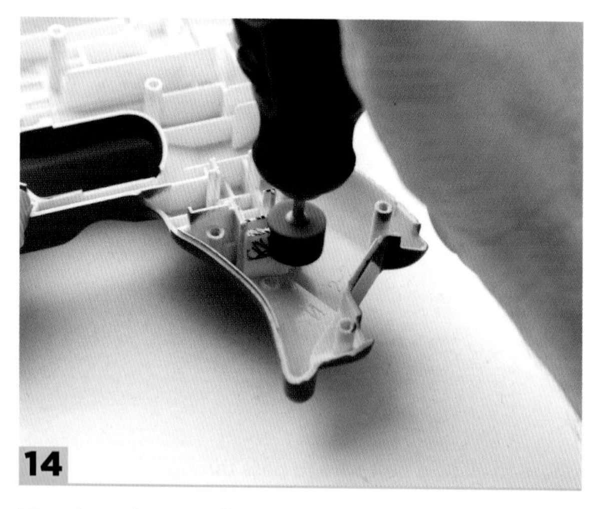
14

Time to make room for an upgraded switch. We'll need to cut or grind down the area marked with a Sharpie.

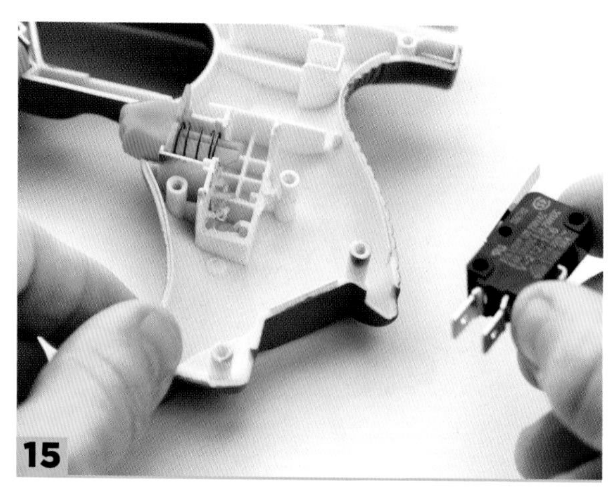
15

Here is what your switch area should like when you're done. That's our new switch on the right, ready to go in!

STEP 3: SWITCH INSTALLATION

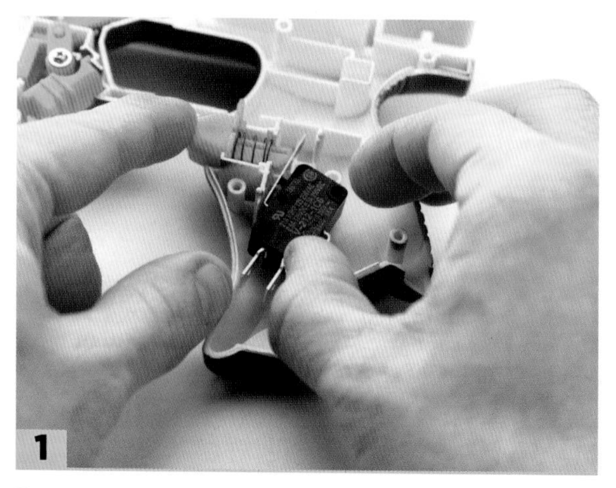
1

Test to make sure the switch fits flat and level in this space. Make sure, too, that the switch's lever arm doesn't touch any plastic and has room to move. When a switch fits properly, there is no looseness or "slop" on the rev trigger, and the switch is properly activated when pressed. Switch placement is perhaps the most challenging part of an electronic Nerf mod. Take your time getting it right.

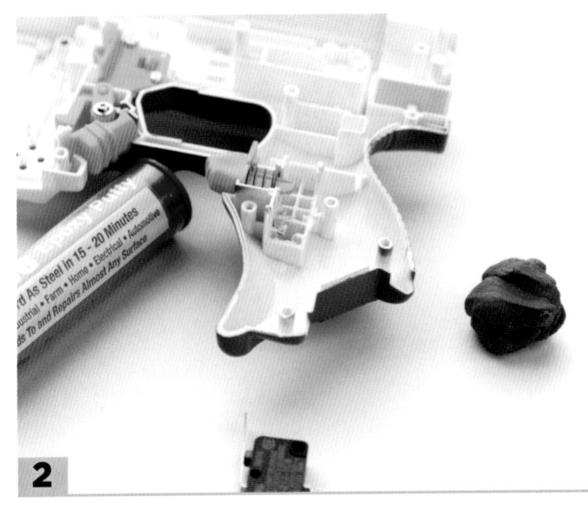

2

The entire area for the new switch must be flat and square. Use epoxy putty or hot glue to create a flat area. Epoxy putty will hold the switch in place like glue. While wearing gloves, knead a small portion of putty together to activate it. Fill the bottom area of the switch as shown. Make sure the switch lines up with the trigger. Use your finger to hold the switch in place and test the trigger. Make sure moving parts don't catch on anything.

> Avoid Super Glue for switches.
> Its vapors can get inside the switch
> and cause it to fail.

3

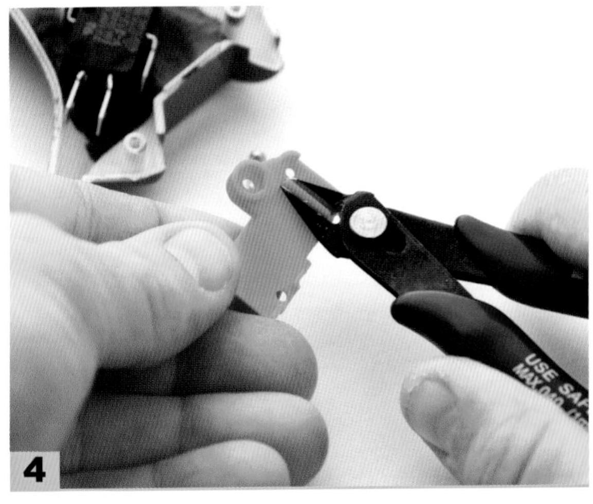

4

Cut the two parts from the switch cover as shown. Screw the switch cover back in place. If your switch is properly placed, the cover should fit.

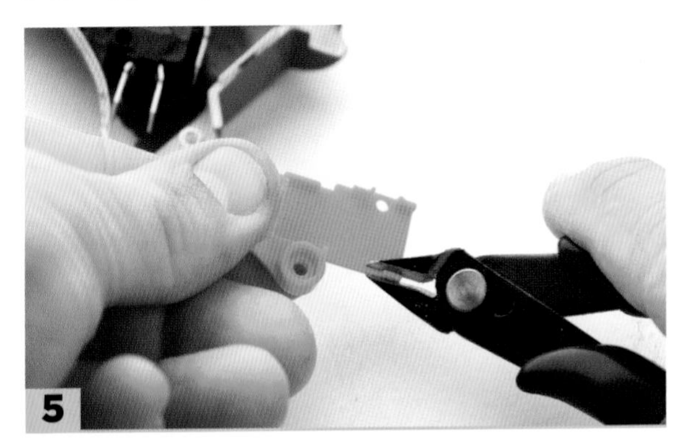

5

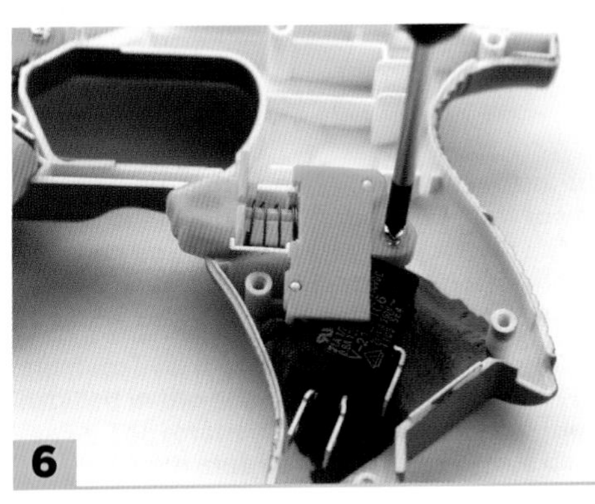

6

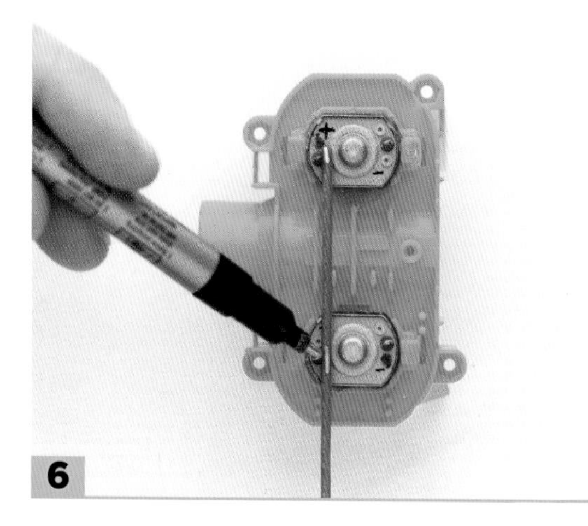

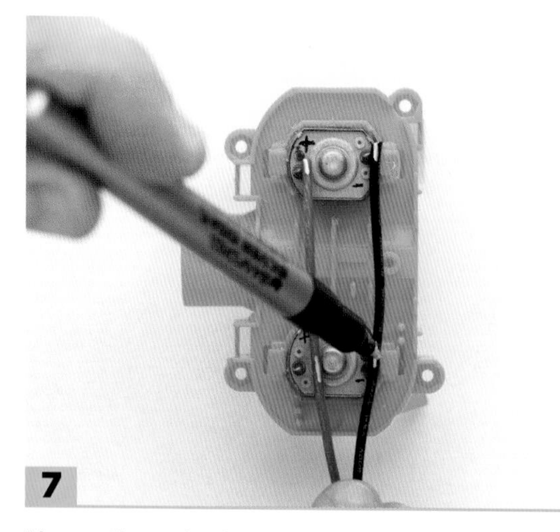

Line up the ends of your black and red wires with the terminals of one motor. Mark each wire positive and negative at the point where it intersects the motor terminals. Your black and red wires could be the reverse of what's here—what's important is that your flywheels spin in the correct directions. Mark these a little long to leave some slack in the wiring between the motors. This will help keep the wiring from pulling on the delicate motor tabs.

STEP 5: SOLDERING

Now it's time for the fun part: making our electrical connections. If you need to, please refer back to Chapter 6 and the advice and tips for soldering. If you're ready, let's get started!

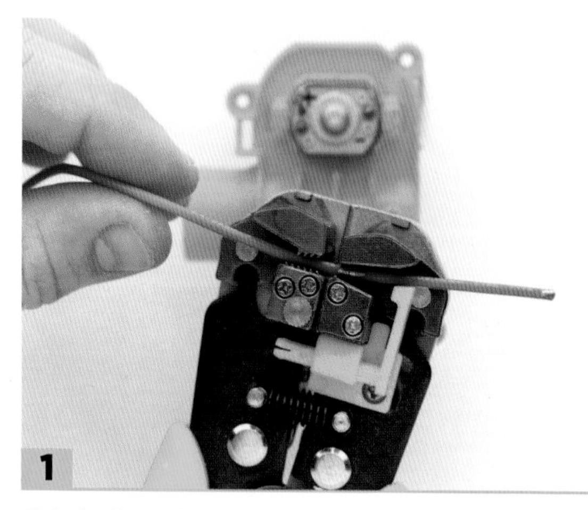

Strip both marked sections of the black and red wire with wire strippers.

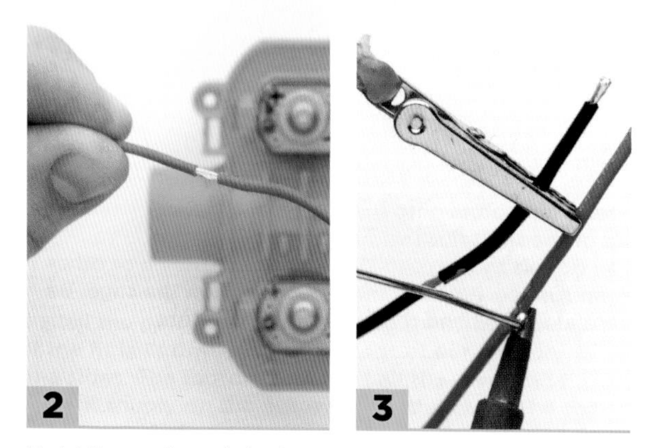

Twist the ends and tin the exposed wire with a bit of solder. This trick allows for one less break in the circuit and makes for a more reliable circuit overall.

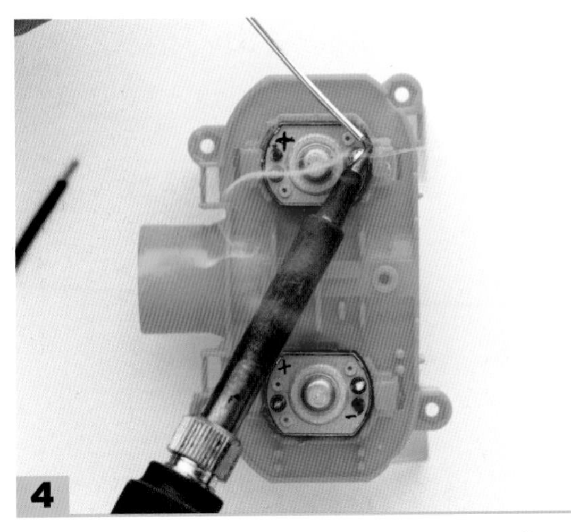

Next, tin the four motor connections. Tinning all connections will make it easier to solder the wires in place.

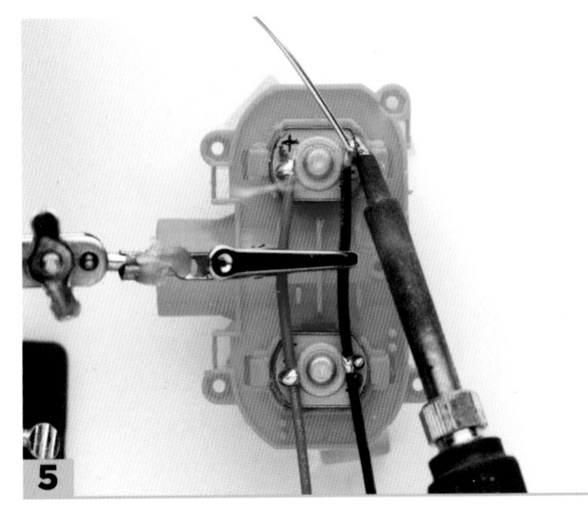

Now we can solder the leads to the motors. Be careful not to use too much solder or allow the solder to touch the motor housing. This can damage the motor and cause a short when you hook up the battery and test the blaster. The small components make this the hardest soldering in a Nerf mod.

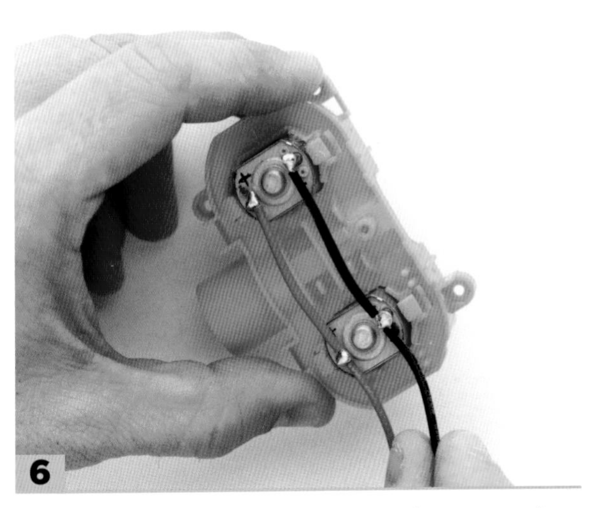

Give the new solder connections a gentle tug to make sure the solder joints are solid.

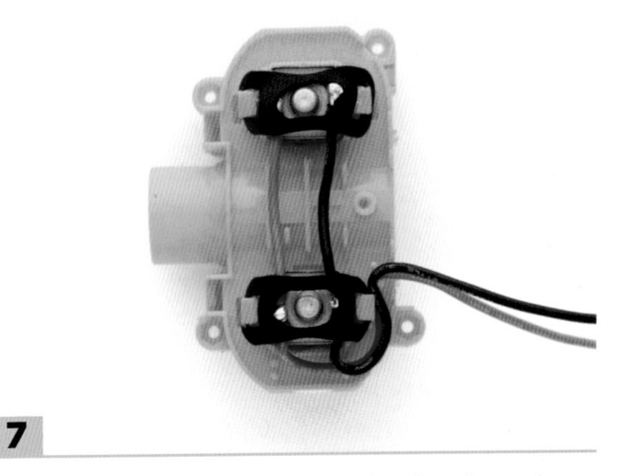

Replace the rubber covers. These absorb noise and lessen strain on the wires. Run the wires as pictured.

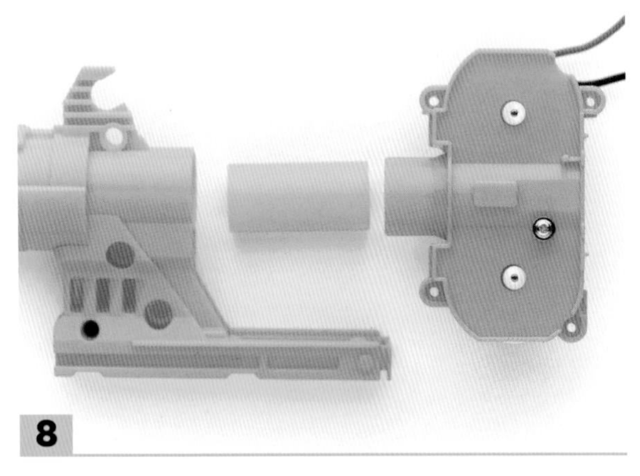

8

Combine the muzzle, barrel, and flywheel cage (left to right) and return them to the blaster.

9

Flip the cage over and line it up with the barrel and muzzle. Place the cage and muzzle back in the blaster. Add the top half of the flywheel cage and fasten the four black screws.

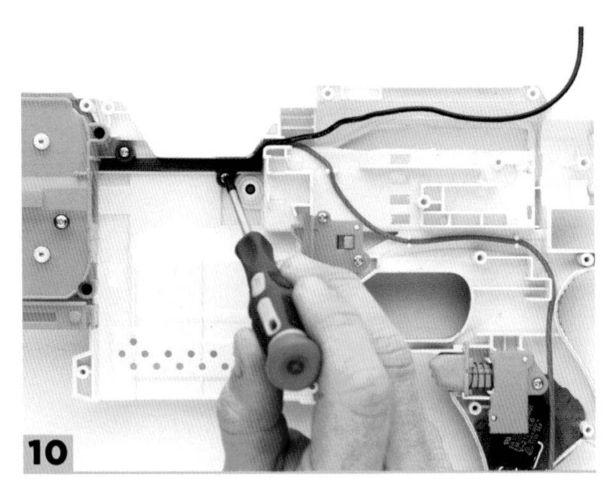

10

Run the wires through their path as shown. Then, replace the black channel guide with the two screws.

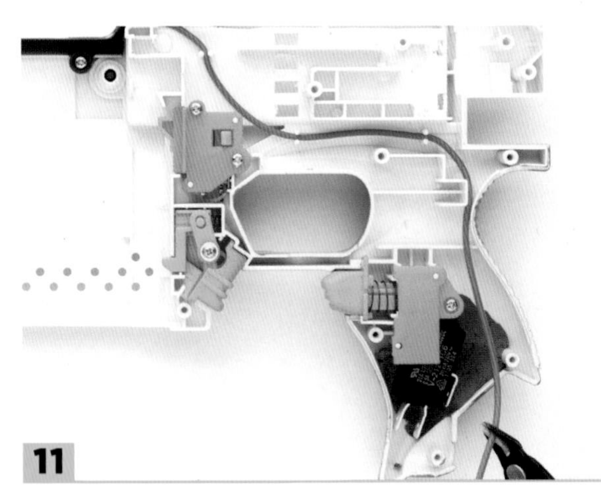

11

Run the red wire through the path as shown. Cut it a little outside the blaster shell to leave some flexibility when soldering it in place.

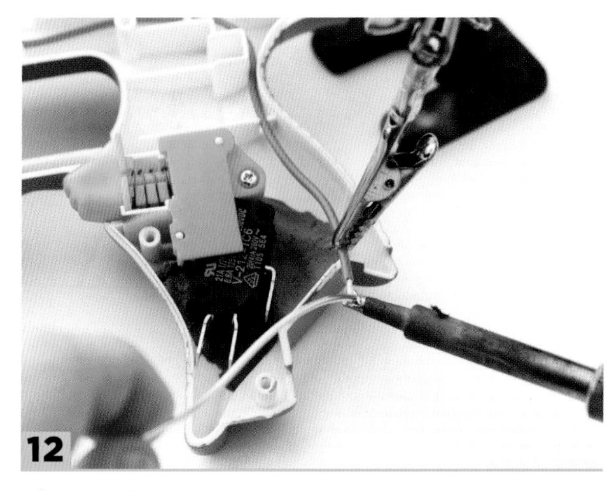

12

Tin the end of the red positive wire to prepare it for soldering.

13

We are going to solder two red wires to the switch terminals marked "NO" for "Normally Open" and "C" for "Common." This means the circuit isn't closed (turned on) until we pull the trigger. On switches like this, these terminals are usually in the same places shown here, but examine your switch to confirm. Note that it doesn't matter which wire you connect to which terminal.

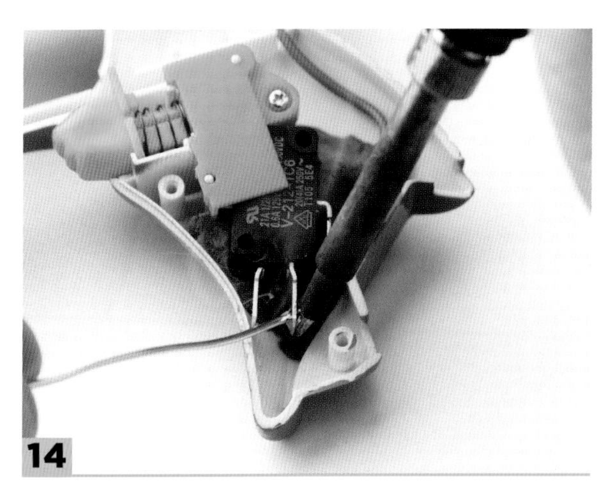

14

Prepare and solder a second piece of red wire to the second NC switch terminal.

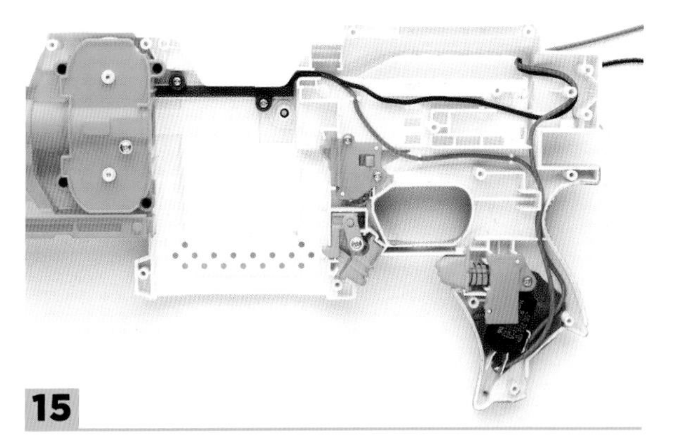

15

Route the second wire you just attached to the switch through the path up the handle as shown

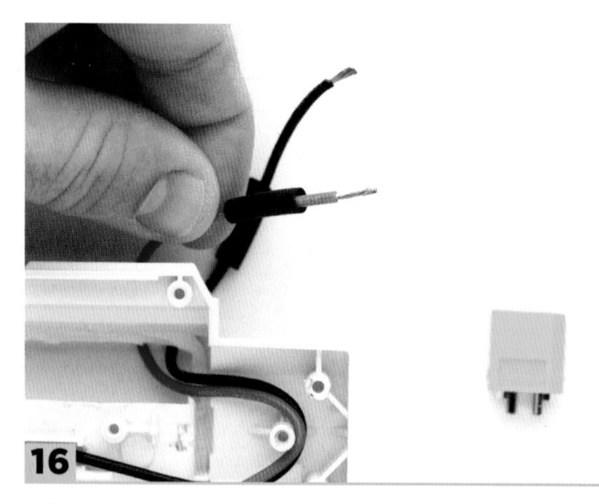

16

Snip the black and red wires at the same point outside the shell. Twist the tips and slide a piece of heat shrink down each wire. Heat shrink prevents the two battery-connection terminals from touching. If they touched, it would short out the circuit and could damage the battery.

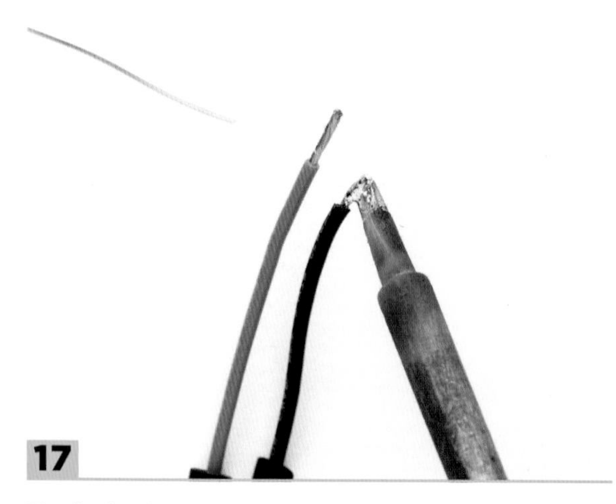

17

Tin the leads of the two wires with a small amount of solder.

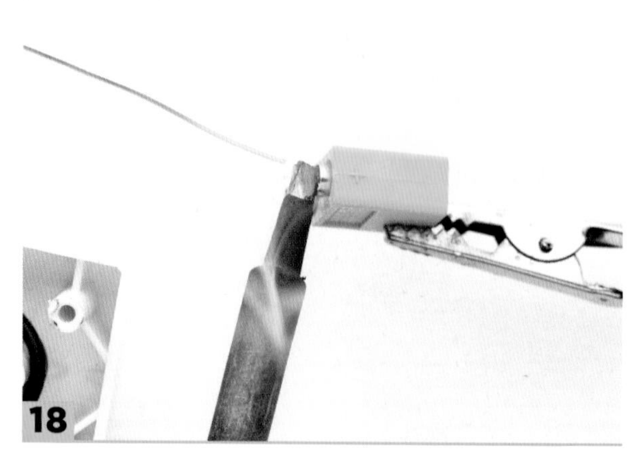

18

Our last step in soldering is to install the XT-60 connector. This will make the blaster compatible with the new battery. Heat up the cup of the XT-60 connector, adding solder to fill the cup.

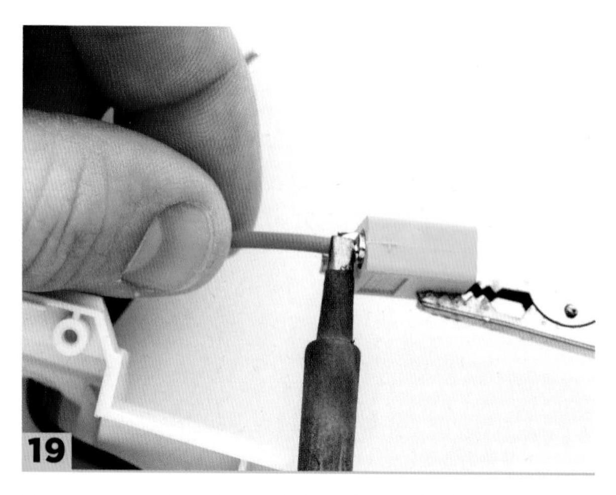

19

Be sure to match the positive wire to the XT-60's positive connection (the flat side is always positive). Insert the wire while heating the XT-60 solder cup and wire at the same time. Hold the wire as still as possible while the solder cools.

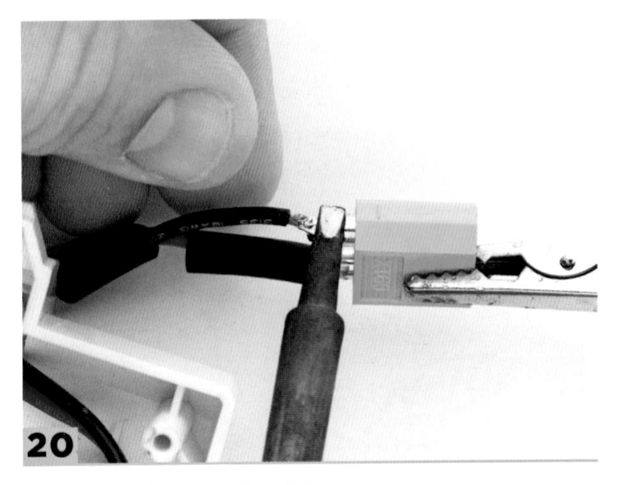
20

Repeat for the negative side.

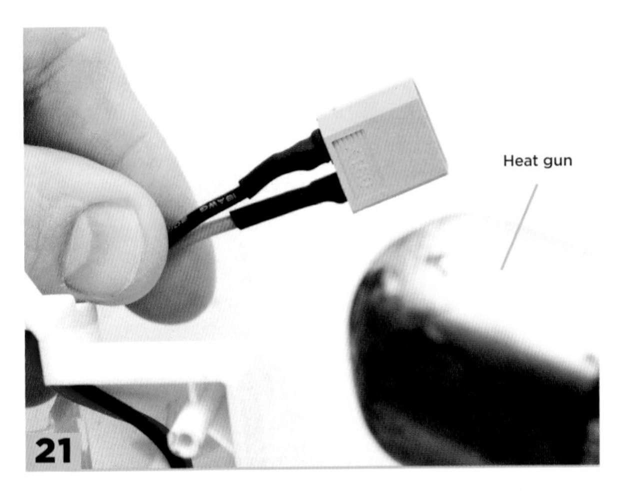

21

Once the joints are cool, slide up the heat shrink to cover the connector contacts. Warm it up with a heat gun, lighter, or candle to shrink the tubing. The tubing shrinks at around 284°F (140°C). The XT-60 connector is nylon and has a melting point of 428°F (220°C).

STEP 6: TEST-FIRING, FINAL CHECKS, AND CLOSING THE BLASTER

We're almost there...just a quick test, and you'll be on your way to impressing your friends with your first successful electric-powered Nerf mod!

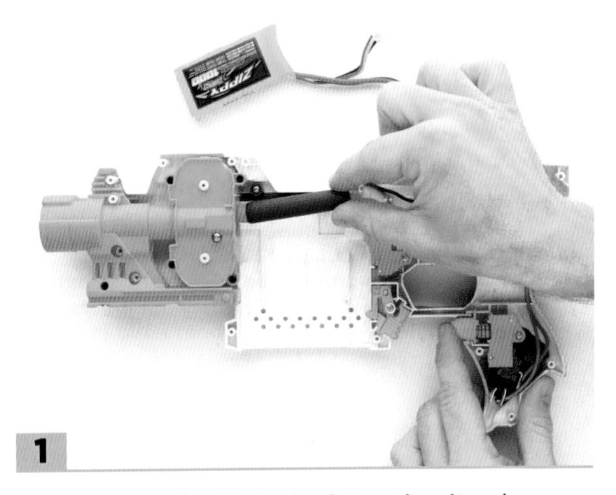
1

Before we put the blaster back together, it makes sense to test it. First, plug in your hot glue gun and set it aside. Hopefully you'll need it in a few minutes. Next, plug in your battery and test your newly wired blaster by pulling the rev trigger and carefully inserting a dart into the flywheels. Make sure the blaster is pointing in a safe direction before doing this. Also, be sure to keep your hands clear of the flywheels.

A wide range of parts I've created. Some of these are just prototypes. Some became final products. You can create virtually anything with 3D printing.

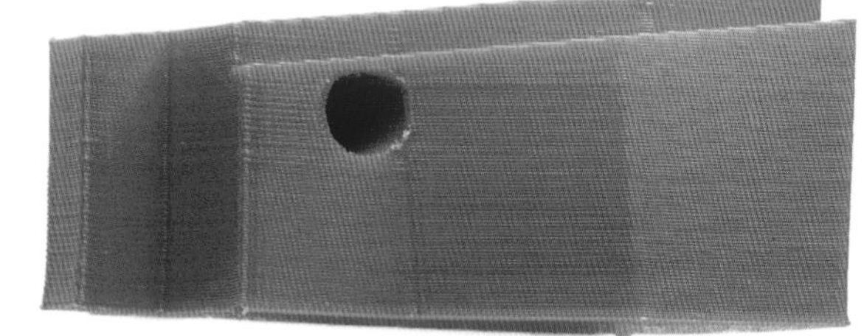

Here you can see the individual layers of a 3D part.

There's no comparison when it comes to the looks of these two blasters. The 3D-printed blaster (above) also performs much better than the PVC tube mod (left). And it weighs less and takes much less time to make.

PRO ADVICE

Jessie from Project Foam Dart Launcher (FDL) says, "Three-D printing has reinvented how we design and create things. It puts small-scale manufacturing into the hands of everyone. In the world of modding Nerf blasters, this means modders can now share their designs with the entire community. It has changed everything."

3D Software

There are many options for 3D design software. I find Autodesk Tinkercad to be a very basic introduction. Autodesk Fusion 360 is an excellent choice for more advanced software. Both are free to download at www.autodesk.com.

Anyone considering a design career should look at Solidworks too. It is the industry standard. I don't use Solidworks because the price starts at $4,000.

Design software can seem difficult, but I learned all my skills in the last three years starting from scratch. I was inspired after seeing the work of Jessie at Project Foam Dart Launcher, an open-source 3D-printed dart blaster.

The FDL-2X (Foam Dart Launcher 2) is a 3D-printed, high-end electronic blaster. It features select fire, brushless motors, variable rate of fire, variable fps, and dart counting. This blaster is simply the highest performing electronic blaster in existence.

The Calliburn is a 3D-printed, spring-powered blaster created by Captain Slug. It is the most accurate blaster I have ever fired. The Calliburn can shoot more than 200 fps. It can also be "tuned." That means different springs can be swapped in for different game types. I run mine with a lower-powered spring for normal games.

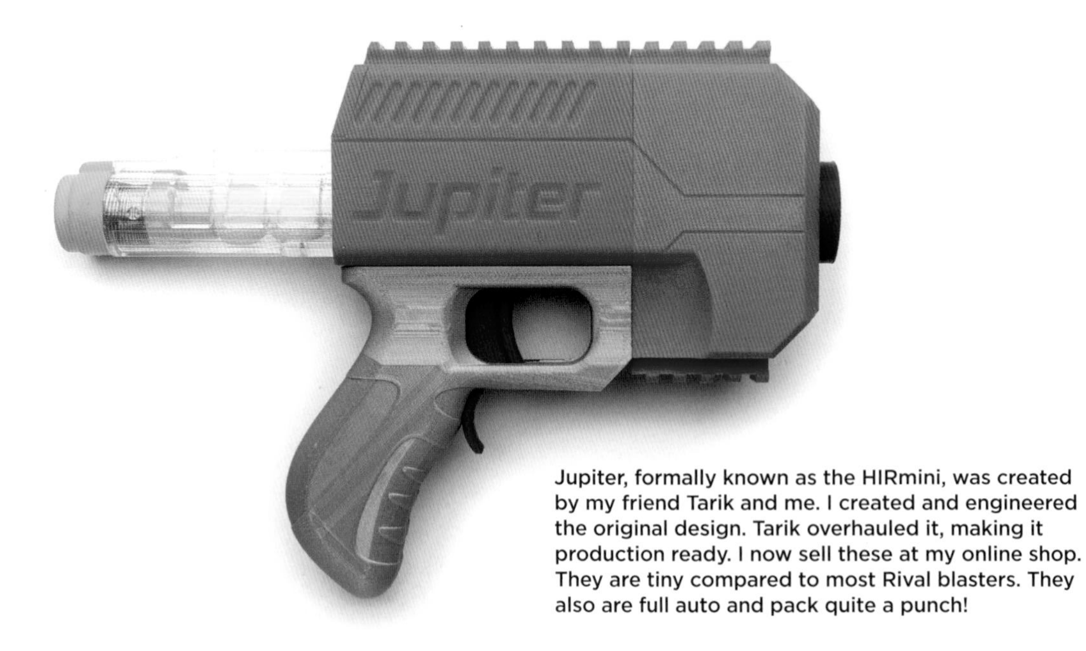

Jupiter, formally known as the HIRmini, was created by my friend Tarik and me. I created and engineered the original design. Tarik overhauled it, making it production ready. I now sell these at my online shop. They are tiny compared to most Rival blasters. They also are full auto and pack quite a punch!

This is the Stinger by Joji at BlasterforgePH. Based in the Philippines, Blasterforge designs tons of completely original blasters from the ground up!

Buying a Printer

I'm frequently asked what printer I recommend. I have owned six different brands of printers. The following two are the only ones I fully recommend. With the fast pace of new releases, I suggest doing some online reading for the most current information. I suspect both of the following companies and printers will be around for years to come.

I highly suggest Makers Muse and Joel the 3D Printing Nerd on YouTube. They are the best sources for 3D-printing tutorials, reviews, and more.

QIDI Tech X-One2

Price: $400

PROS

This is a great entry-level printer. It prints very well and is very reliable. I own eight of them, and I've run each of them for thousands of hours. The company has excellent customer support despite being based in China. They have responded to my questions and warranty requests within hours. QIDI also makes several other models, including the X-Smart and X-Pro.

CONS

Small print surface. This printer is best for smaller parts.

Prusa i3 MK3

Price: $1,000 assembled; $750 DIY kit

PROS

This printer has every bell and whistle a hobby or professional user could want. There is also a huge community of users because it is an open-source business model. There are third-party versions that cost much less than the Prusa. You can even build your own. In my opinion, the Prusa is still the best one out there. Beware of poorly produced clones and always read current reviews for the best advice.

The company even prints these printers *with* their printers (all orange plastic parts in the photo are 3D printed). As far as I'm concerned, this is the best printer under $2,500.

CONS

Because the company is based in Prague, Czech Republic, getting the parts and printers takes time. The company is growing fast, but customer service is not as prompt as with other companies.

Kinds of Paint Jobs

There are many kinds of paint jobs. They range greatly in difficulty and cost. In this chapter, we will look at the most basic sort of paint job. It's the same process I used for my first paint job.

This humans vs. zombie-themed blaster was painted by Tom at Foam Data Services. This is some seriously high-end painting. Tom's work is second to none in the hobby. *Photo by Tom FDS Rayven*

This blaster is called the "Copperhead" and was painted by N-E-R-M at Northern Bay Area Coalition of Nerf in California. It includes hydrographic film with a sweet snakeskin pattern. N-E-R-M says, "I walk a fine line between passion and obsession." *Photo by N-E-R-M*

Spray-on (easiest): These are sometimes called "rattle-can" paint jobs. That's because you use aerosol spray paint with a rattling ball inside. The ball helps mix the paint.

Spray-on paint jobs involve four steps: sanding the blaster, spraying primer on the blaster, painting the blaster, and, finally, spraying a clear coat on the blaster to protect your new finish. We'll look at each of these steps in this chapter's project.

Paint cans can range from $2 for basic paint to $20 for automotive-grade paint. Usually, anything that can paint on plastics can be used. Read the can's label. I use an automotive paint in this chapter that costs around $7 a can.

Hydro-dipping (moderate): Hydro-dipping involves using a patterned hydrographic film. This thin film is floated on water. The shell is carefully dipped in the water. An activator makes the film stick to the blaster. Hydro-dipping can have stunning results.

Swirl painting (moderate): In this method, oil-based paint is floated on water. The paint is swirled with a brush or stick to create interesting patterns. Then the shell is dipped in the liquid. The results can be striking, as seen in the blue and white Hammershot on page 113.

Airbrushing (difficult): Airbrushing can be used for basic paint jobs and for more advanced details on blasters. Airbrushing refers to the tools used and not the paint itself. An airbrush uses an air compressor to spray the paint. A compressor applies paint more evenly than rattle cans do. The result is a more professional paint job. Cars are usually painted with an airbrush. Airbrush paint bottles are smaller than rattle cans. You can usually find a wider variety of colors too.

Hand painting (difficult): This method is difficult to do well—and easy to do poorly. Hand painting is exactly what it says: using brushes, sponges, or another hand tool to paint your blaster. This method can be a lot of fun. It can also be used to add small details to a paint scheme, as we'll see in this chapter's project.

PAINT SAFETY

Always paint in a well-ventilated area. This means an area with good airflow. An open garage or outdoors is best. Always wear a respirator. Your respirator should have a cartridge and be designed specifically for use around paint fumes. Always read instructions on paint cans or bottles—they will have the best information for safety and application.

RATTLE-CAN PAINT JOB

Time: 1–3 hours for prep; 30–60 minutes for painting; up to a few weeks for drying between coats, depending on humidity and temperature

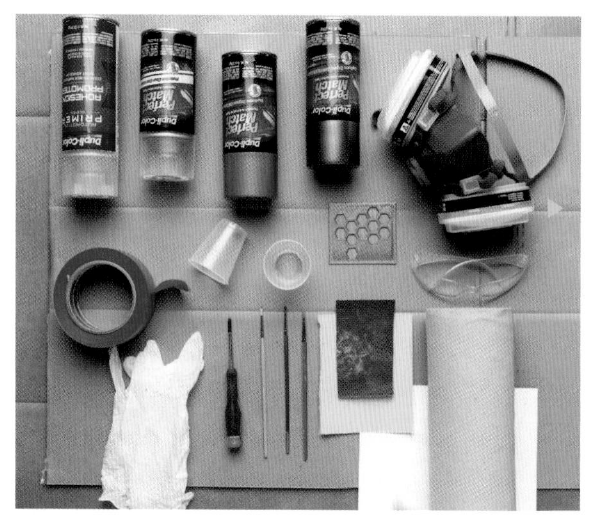

Hardware stores and big-box stores, such as Lowe's, carry some spray paints that work on plastic. You don't have to use high-grade paint like the one used here. Choose a paint that fits your budget. But be ready to sand a lot more if you use primer instead of adhesion promotor.

PAINTING TECHNIQUES USED

- Sanding
- Masking
- Applying primer or adhesion promote, paint, and clear coat
- Stenciling to add details
- Hand painting additional detail areas

TOOLS AND MATERIALS

- Paper and pencil
- Screwdriver
- Gloves
- 400- and 800-grit sandpaper
- Putty knife (optional)
- Dish soap
- Painter's tape
- Cardboard for painting on and controlling overspray
- Respirator
- Primer or adhesion promoter
- Spray paint
- Stencil (optional)
- Paintbrush
- Clear-coat spray

The best temperature for painting is between 50° and 80°F (10° and 27°C) with less than 75 percent relative humidity. Humidity is the amount of moisture in the air. When painting, humidity from 40 to 50 percent is best. Higher humidity can cause trouble with the paint as it dries.

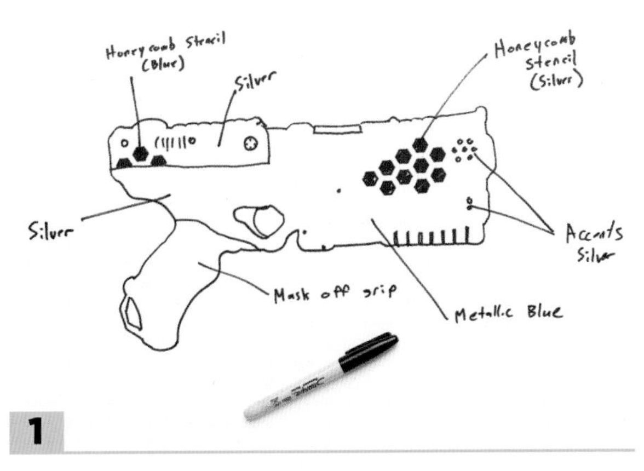

Start by planning your paint job. I'm going to keep my blaster handle black and paint the rest light metallic blue. I'm going to paint the top slide metallic gray. Then I plan to use a 3D-printed stencil to add honeycomb patterns on the slide and the front of the blaster. I roughly traced the blaster on a sheet of paper and drew in the details by hand. My drawing will remind me of the order the steps need to be completed in.

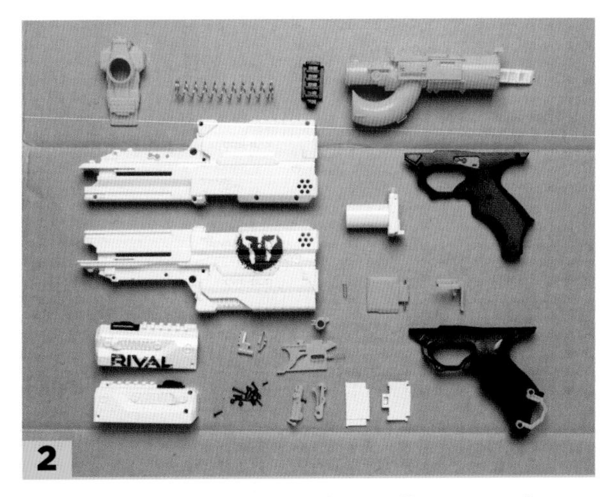

Before painting the blaster, unfasten all screws and remove as many internal parts as possible. I suggest painting the two shells separately. If you paint the shells while they are attached to each other, you'll have a hard time avoiding blemishes. Also, the paint can act like glue and cause the shells to stick together. I also recommend leaving all orange parts orange for safety reasons.

Wearing gloves, sand both shells. This can take a bit of work. But the better you sand, the better the paint will stick to the blaster. If you use adhesion promoter instead of primer, you can skip this step almost completely. Just remember to clean the blaster with dish soap and allow it to dry. This removes oils and dirt before painting.

The small grooves can be difficult to sand. Sometimes a putty knife can help get the sandpaper into the cracks.

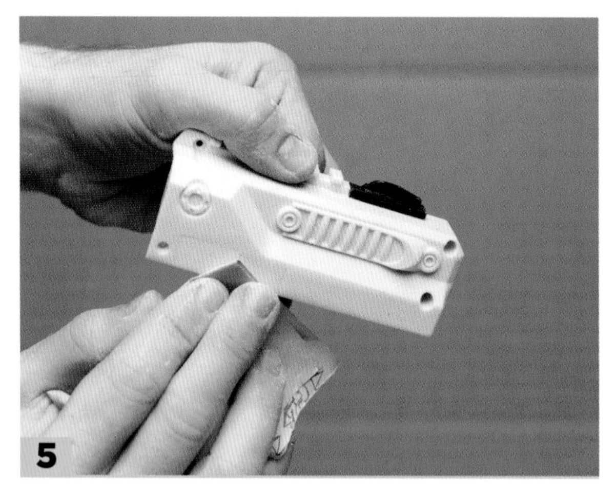

Use 400-grit sandpaper to remove logos for a cleaner look. Work up to 800-grit to smooth the surfaces. The smoothness of your sanding job will determine the final paint quality. Paint will not hide holes and scratches.

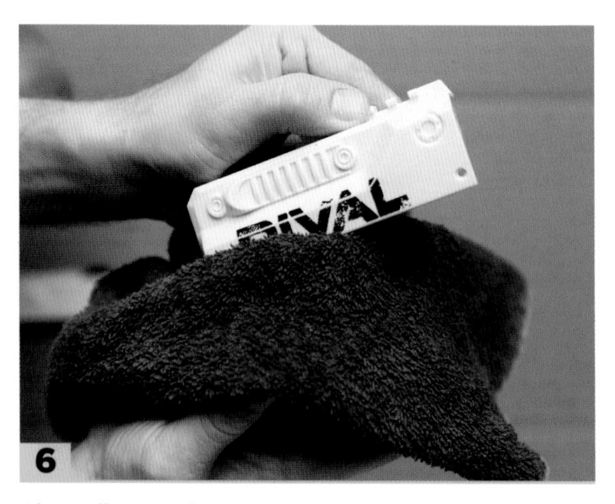

Clean all parts after sanding to remove any dust or debris. You can use dish soap and warm water to clean parts. Just make sure they are totally dry before you start to paint.

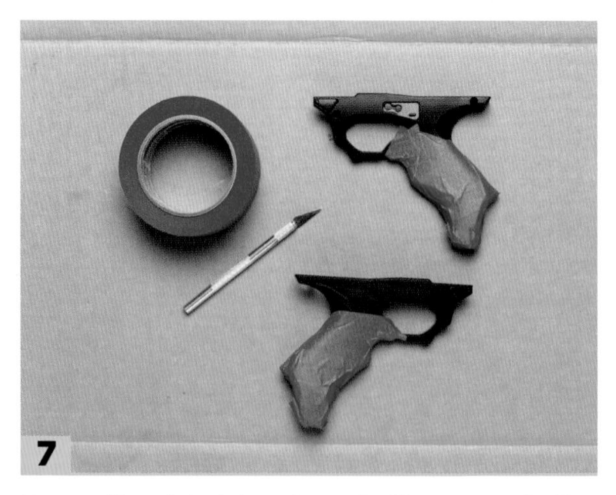

Use quality painter's tape to mask off areas you don't want painted. The blaster handle receives the most wear. When the color works with my design, I like to keep it unpainted.

Trim the tape for sharp edges that follow the lines of the blaster. You don't have to mask off your handle, but this shows what you can do with masking.

9

Find a well-ventilated area and lay down old cardboard, paper, or plastic to catch overspray. Place the shells on upside-down boxes. This will allow spray to reach all sides of each shell. I paint in my garage with the large doors completely open. Remember to wear a respirator! And before starting, read the labels on your primer and paint cans for information about drying times.

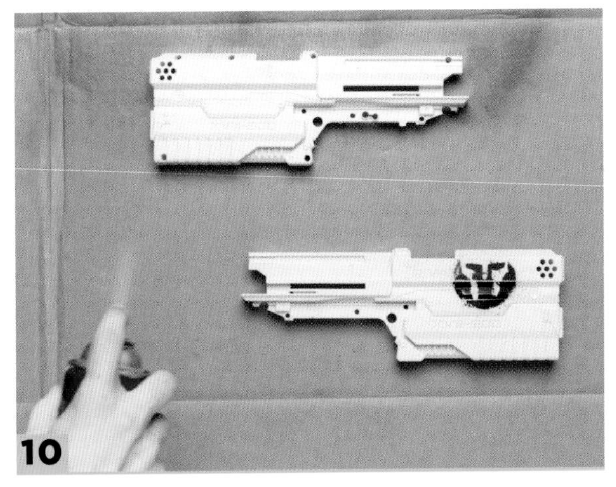

10

Spray on the primer. Press the nozzle all the way down and use short strokes. Always start with the nozzle pointed off the blaster and then guide it over the blaster. Use just enough paint to cover the shell, but not so much that it pools. Several thin layers of paint will always look better than one thick layer. My paint says to apply two or three layers, so I'm going to do three. You want to apply primer from multiple directions to get good coverage.

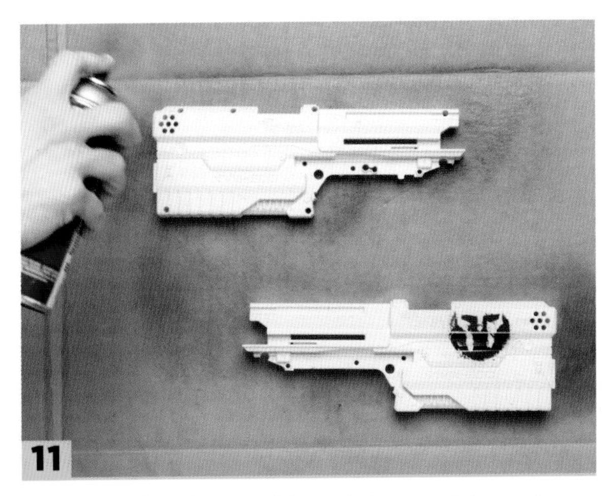

11

Always apply primer, paint, and clear coat from all directions. This will help you reach down into all the small grooves on the blaster shells. The process for applying primer, paint, and clear coat is the same. But read your cans: the drying time for each might be different.

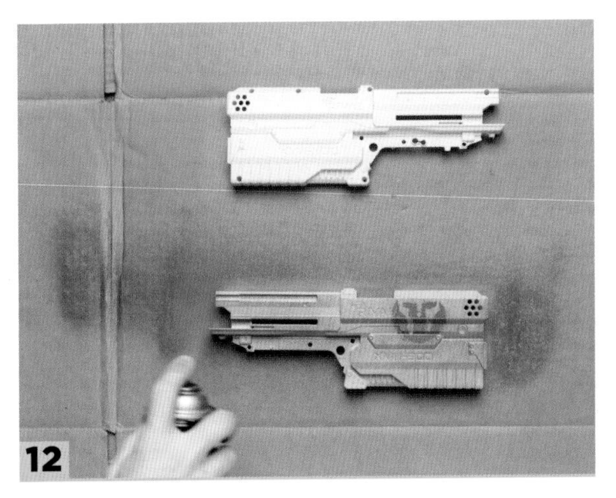

12

When the primer or adhesion promoter is dry, spray on your first layer of paint. You want to apply paint thinly at first. Start spraying before the nozzle is pointed at the blaster. Again, work your way around and spray from multiple angles. I chose light-blue automotive paint. Then wait for this to dry.

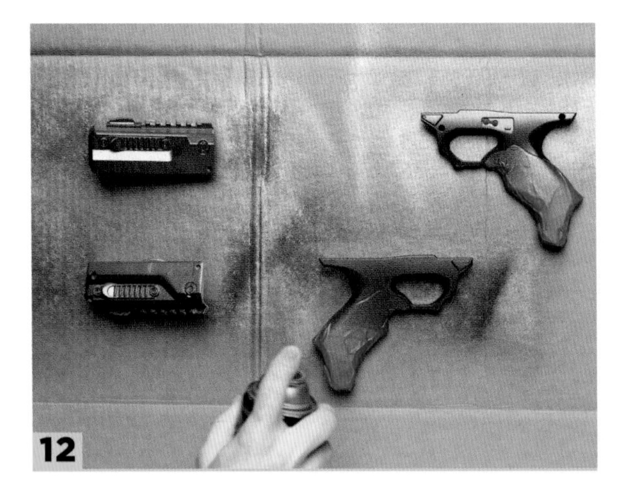

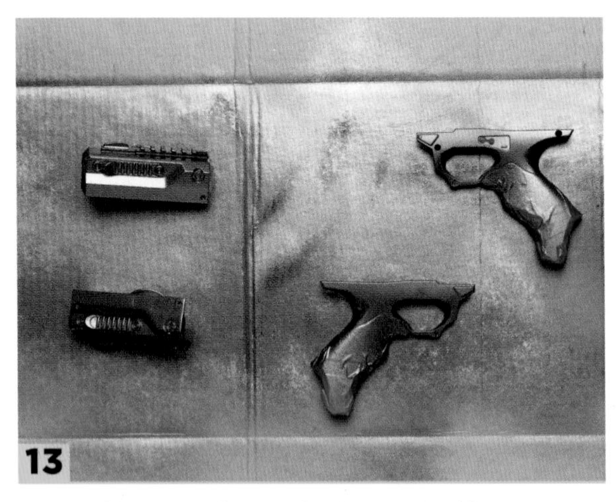

Next, add a second coat of paint to your blaster. Remember, a few thin coats are better than one thick coat. Use too much paint and it will probably pool and run. Repeat this layer for each part. Wait for it to dry, and spray a third coat if needed. Here I'm painting the handle and top slide in metallic silver.

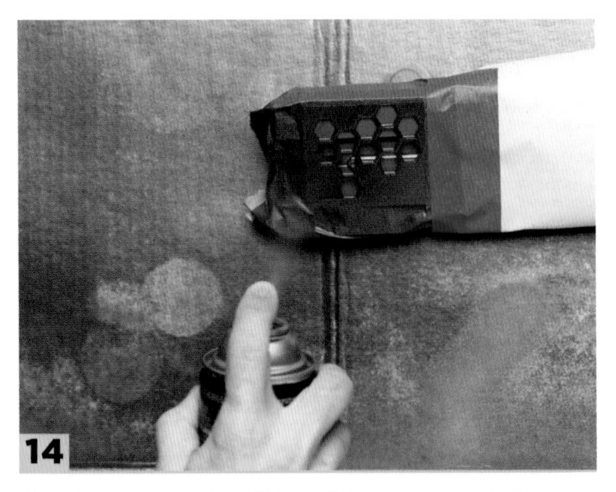

Here, I'm using a stencil to paint on an accent. The stencil is a simple 3D-printed part that I designed to add some detail to the paint job. Many stencils are available online. Or you can make your own from various objects. Some store-bought stencils have a sticker-like adhesive for extra-clean results.

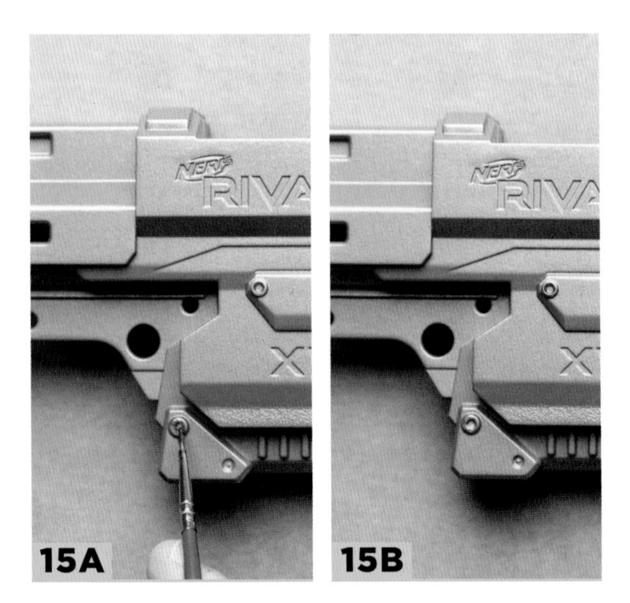

When all the spray painting is complete, I apply some details with a paintbrush.

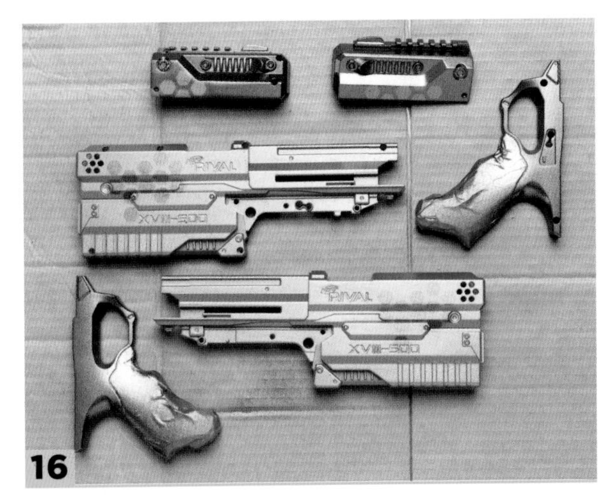

16

With the detail work done, just one step remains: clear coat!

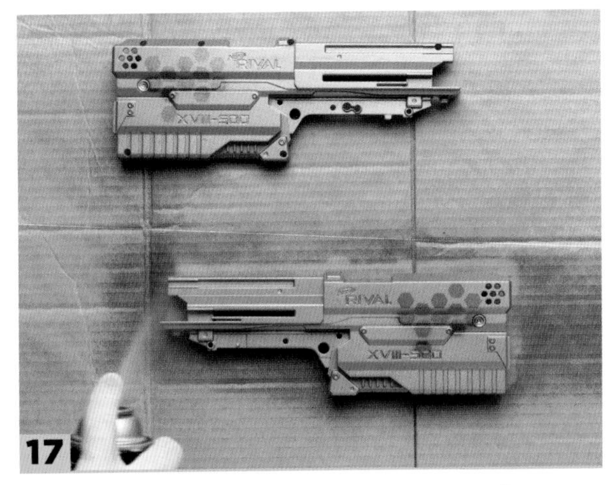

17

Clear coat protects paint from UV rays, scratches, chips, and general wear. The more thin layers you apply, the smoother your finish will be.

18

Now it's time to be patient. Wait for your clear coat to dry. Make sure you give it plenty of time. I generally wait twice as long as the can recommends. If your blaster is at all sticky, it needs more time to dry. When it's completely dry, you can remove any painter's tape.

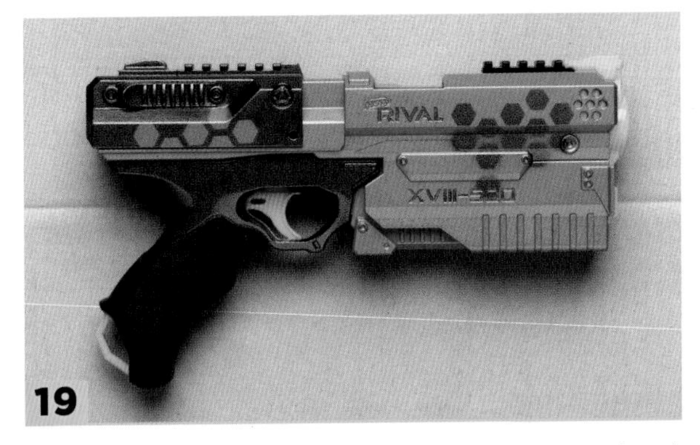

19

Congrats! If you've followed along, you have completed your first paint job. Now you can reassemble your blaster. Take care not to scratch your new paint! I'm pretty happy with how this one turned out. Every time I paint I learn something new and get a bit better. Remember, it takes practice—you won't get it perfect on your first paint job. And that's okay.

QUICK TIP

While photographing this paint job, my dad gave me a great piece of advice. After you finish painting with one can, clear the nozzle by spraying it while holding the can upside down. This helps the pressurized gas in the can clear any paint from the nozzle. Now the next time you pick up the paint, it will still be usable!

Combination

WalcomS7 chimes in, "What makes any blaster better? Having two of them." WalcomS7's "Deleter" is the fastest-firing Nerf pistol in existence, emptying 36 darts in one second. The HyperFire was chosen because it's thinner than other full-auto Nerf blasters.

Side view

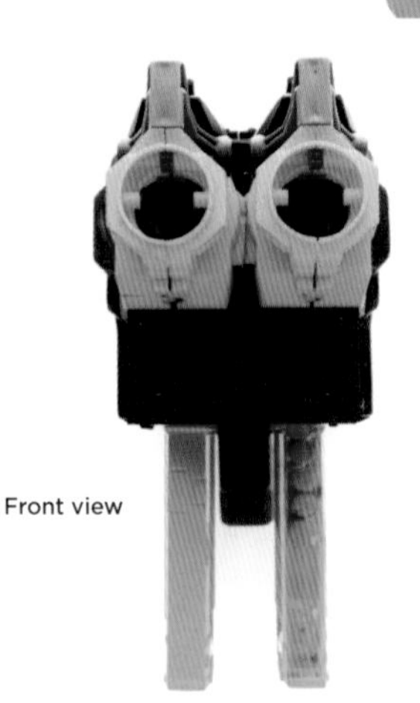

Front view

This incredible blaster is two Nerf HyperFire blasters combined to make what WalcomS7 calls the "Deleter." It's beautifully painted as well.

The Poonbow is a beautiful integration of a Nerf Stryfe and a vintage Nerf Crossbow. *Photo by Coop772*

The PwnSaw blaster from YouTuber IamBobololo is another top-tier mod. It features a Stampede integrated with a RapidStrike along with a SwarmFire connected up front.

Hopper Extensions

One relatively simple mod is to extend the hopper capacity of a Rival blaster. The Rival Nemesis, Dart Zone Powerball, Rival Prometheus, and Adventure Force Accelerator all feature hoppers that can be easily extended. Sometimes this is done with 3D printing, as with the Nemesis funnel shown here. Other times it is done by simply gluing a container in place (such as one of those giant cheeseball containers from a warehouse store).

This simple funnel designed by Philip at the Open Flywheel Project makes it easier to load the Nemesis.

Humorous Mods

Sometimes it's fun to do something just for laughs. My friend Brett made this SaxBZ: an XBZ (Buzz Bee Extreme Blastzooka) air-pressure tank inside a toy saxophone. He spent an entire season of games trying to tag every player out with this one.

The SaxBZ fires two missiles. My friend Brett even managed to tag me with it at a game.

RESOURCES

YouTube Channels

Lord Draconical
www.YouTube.com/lorddraconical

Drac has a wide range of mod-related Nerf content. He posts several videos a week.

Foam Data Services (FDS)
www.YouTube.com/foamdataservices

Tom at FDS is a top-notch painter and very technical modder.

IamBobololo
www.YouTube.com/iambobololo

Bobo's sense of humor alone is reason enough to watch his channel. He does original mods, reviews, and a lot more fun stuff!

Jangular
www.YouTube.com/jangular

Jangular has a great weekly wrap-up of Nerf news: releases, blaster leaks, new mods, and a video of the week. If I watch just one Nerf-related thing each week, it's his weekly wrap-up.

Make Test Battle
www.YouTube.com/maketestbattle

Ryan, Alex, and Justin run this channel from Australia. They also produce a lot of Nerf-specific motors, which many retailers in the United States carry. Their content is very unique and well produced.

Captain Xavier
www.YouTube.com/captainxavier

Captain Xavier continues to pump out tons of useful Nerf mod content.

OutofDarts
www.YouTube.com/outofdarts

This is where my video content lives. I have worked for the last 13 years as a cinematographer and filmmaker, so my goal is to produce some of the best-looking content out there.

Favorite Shops

Local Shops. Nerf is a frustrating hobby to shop for locally. You can always find blasters at major big-box stores such as Target and Walmart.

R/C hobby shops carry some basics such as batteries, wire, and connectors. You'll often find they don't have quite the right size battery packs for Nerf.

Thrift stores are great places for inexpensive blasters. A large portion of the Nerf community relentlessly searches for these bargains. Thrifting can be a ton of fun.

Out of Darts (www.outofdarts.com). This is a shameless plug for my own shop. When I first joined this hobby, I was frustrated that I had to order parts from five or six different places for one simple electronic mod—while paying shipping at each store! I carry nearly everything a Nerfer could want: supplies, parts, 3D designs, tools, accessories, and even completed blasters.

FoamBlast (www.foamblastshop.com). Makers of motors, cool 3D parts, and more; I like their stuff so much I sell some of it myself. Michelle and Adrianna are wonderful to work with and have a great reputation for customer service.

Containment Crew (www.containmentcrew.com). Based out of New York state, Containment Crew produces flywheels and cages, and carries a large selection of nerf mod parts, accessories, and supplies. They are also just about the nicest guys you'll meet.

Blasterparts (www.blasterparts.de). Likely the largest manufacturer of Nerf-specific parts outside of China, Blasterparts, based in Berlin, Germany, produces quality products. Many of their products can be found in the United States, both at Containment Crew and Out of Darts.

Website

Blaster Tag Association

www.blastertagassociation.com

BTA is a competitive Nerf league with games around the world. They feature a variety of game types. You can even start your own league and join the association.

Nerfers Online

www.Reddit.com/r/Nerf

www.NerfHaven.com

Nerf Modders Welcome group on Facebook

ACKNOWLEDGMENTS

I want to thank everyone who has helped make this book possible. First, to my wife, Pamela, for never doubting this obsession for a minute. To Mom and Dad, thank you for inspiring me to get interested in making things, buying me my first R/C car (on which I learned to solder), and supporting me during the final leg of completing this book. Thank you to my dad for his editorial feedback on this book. Thank you to my sister for her tireless work in the Out of Darts shop while I was writing this book and welcoming my daughter to the world. To John and Alex, thank you for the help getting the shop off the ground.

Thank you, also, to the amazing Nerf community for all the love, support, and help in creating this book. This book is aimed at bringing more awesome modders into our community. I hope it's a helpful starting place in the Wild West of Nerf mods!

Thanks to Greg and Rachel for their great painting advice.

There are so many YouTube personalities and friends who pulled me into this hobby. Thank you for creating such fantastic content and sharing it with us all: Lord Draconical, Coop772, IAmBobololo, Tom at FDS, Jangular, Make Test Battle, PDK Films, Captain Xavier, WalcomS7, Mag212, Zombona Machine, Bay Area Nerf, FoamBlast, North Bay Nerf, Project FDL, Beret, and many more.

To the wonderful groups I've had the chance to play with, keep on flinging foam: NoBaCon, East Bay Area Nerf, Bay Area Urban Recreation Nerf (B.U.R.N.), Pacific Northwest Nerf Club (PaNNC), Southbay All-Ages Nerf Group (SAANG), and Endwar, to name a few.

Finally, a huge thanks to everyone watching my channel, supporting my shop, and reading this book. Happy modding!

You can find me at:
Website: www.outofdarts.com
YouTube: www.youtube.com/outofdarts
Instagram: @outofdarts

ABOUT THE AUTHOR

Luke Goodman is an avid Nerf enthusiast better known as YouTuber "Out of Darts." He has always loved building, tinkering, and learning how things worked.

For the last thirteen years, Luke has worked as a self-employed corporate and commercial filmmaker after graduating with a BFA in cinematography from the Academy of Art University in San Francisco. His love of filmmaking collided with the Nerf hobby, and his YouTube videos focus on new original creations, modifications, tutorials, and other all-ages Nerf content. He also owns an online Nerf mod hobby shop.

Originally from Minnesota, Luke currently lives in Washington State with his wife, Pamela, daughter, Hazel, and dog, Buko. When not thinking up new designs and ideas, he spends his free time outdoors in the Pacific Northwest, gardening, hiking, and camping. He also loves swimming and music and enjoys playing the piano.

INDEX

Page numbers in **bold** refer to illustrations in the text.

Brimming with creative inspiration, how-to projects, and useful information to enrich your everyday life, Quarto Knows is a favorite destination for those pursuing their interests and passions. Visit our site and dig deeper with our books into your area of interest: Quarto Creates, Quarto Cooks, Quarto Homes, Quarto Lives, Quarto Drives, Quarto Explores, Quarto Gifts, or Quarto Kids.

First published in 2018 by Voyageur Press, an imprint of The Quarto Group,
401 Second Avenue North, Suite 310, Minneapolis, MN 55401 USA.
T (612) 344-8100 F (612) 344-8692 www.QuartoKnows.com

Voyageur Press titles are also available at discount for retail, wholesale, promotional, and bulk purchase. For details, contact the Special Sales Manager by email at specialsales@quarto.com or by mail at The Quarto Group, Attn: Special Sales Manager, 401 Second Avenue North, Suite 310, Minneapolis, MN 55401 USA.

10 9 8 7 6 5 4 3 2 1

ISBN: 978-0-7603-5782-8

Library of Congress Cataloging-in-Publication Data

Names: Goodman, Luke, 1984- author.
Title: The NERF Blaster modification guide : the unofficial handbook for making your foam arsenal even more awesome / Luke Goodman.
Description: Minneapolis, Minnesota : Voyageur Press, an imprint of The Quarto Group, 2018. | Includes index.
Identifiers: LCCN 2018017417 | ISBN 9780760357828 (flexi-bind)
Subjects: LCSH: NERF toys.
Classification: LCC GV1220.9 .G66 2018 | DDC 790.1/33--dc23
LC record available at https://lccn.loc.gov/2018017417

Acquiring Editor: Dennis Pernu
Project Manager: Alyssa Bluhm
Art Director: Cindy Samargia Laun
Cover and Page Design: Foltz Design, LLC

Printed in China